Basic Carpentry Techniques

Created and Designed by the Editorial Staff of Ortho Books

Project Editor
David W. Toht

Writer
Roger S. Grizzle

Illustrator
Mario Ferro

Ortho Books

Publisher
Robert B. Loperena

Editorial Director
Christine Jordan

Managing Editor
Sally W. Smith

Acquisitions Editors
Robert J. Beckstrom
Michael D. Smith

Prepress Supervisor
Linda M. Bouchard

Sales & Marketing Manager
David C. José

Publisher's Assistant
Joni Christiansen

Graphics Coordinator
Sally J. French

Editorial Coordinator
Cass Dempsey

Copyeditor
Judith Dunham

Proofreader
Alice Mace Nakanishi

Indexer
Frances Bowles

Separations by
Color Tech Corp.

Printed in the USA by
Banta Book Group

Thanks to
Deborah Cowder
Jef Flanders
Hank Hanson, Future Fence
Kaufman and Broad—South
  Bay, Inc.
David Van Ness

Builders
Names of builders are fol-
lowed by the page numbers
on which their work appears.
L=left; C=center; R=right;
  T=top; B=bottom

Meadowcreek by Radisson
  Development Corp.: 24,
  50–51, 90BL, 90BR
Silverhawk & Company, Inc.:
  71, 90TR

Photographers
Names of photographers are
followed by the page numbers
on which their work appears.
L=left; C=center; R=right;
  T=top; B=bottom

Michael Landis: 17, 24, 50–51,
  90CL, 90CR, 91C, 91B
Fred Lyon: 3B, 7, 22–23
Geoff Nilsen: Front cover
Ortho Photo Library: 71, 90TL,
  90TR, 90BL, 90BR, 91T
Chriss Poulsen: Back cover
Kenneth Rice: 49
Gary Russ: 1
Kevin Vandivier: 3T, 4–5

Address all inquiries to:
Ortho Books
Box 5006
San Ramon, CA 94583-0906

© 1981, 1997 Monsanto Company
All rights reserved

1	2	3	4	5	6	7	8	9
97	98	99	2000	01	02			

ISBN 0-89721-319-X
Library of Congress Catalog Card
Number 96-67948

THE SOLARIS GROUP
2527 Camino Ramon
San Ramon, CA 94583-0906

Front Cover
Today's carpenter relies on
a broad range of tools, from
simple hand tools that have
not changed in their basic
design for many generations
to power and pneumatic tools
with increasingly sophisticat-
ed designs. Whether you work
with a treasured collection of
hand-me-downs or the latest
models from the showroom
floor, you will want to orga-
nize them with a toolbox similar to the one shown here (for
another view, see page 24; to
build the box, see page 40). A
detailed discussion of carpentry tools begins on page 23.

Title Page
This carpenter is toenailing
a 2×8 valley jack rafter to the
2×10 ridge board. Building
codes specifiy the number and
size of nails required for such
connections, and the maximum length that various sizes
of lumber may span when
used in specific applications,
such as rafters. Framing a
complex roof is beyond the
scope of this book, but you
will find techniques for framing a simple gable roof beginning on page 72.

Page 3
Top: Temporary diagonal
braces hold these walls in
place until permanent bracing
or sheathing can be installed.

Bottom: Tools needed for
measuring, marking, cutting,
positioning, and fastening
boards are all in this tool belt.
Having them close at hand
prevents wasted effort looking
for them around the job site.

Back Cover
There is something very
satisfying about standing back
to view your project at the
end of the day and seeing
what you have accomplished.

Basic Carpentry Techniques

GETTING STARTED

Whether you are putting an extra shelf in a closet or building a house from the ground up, the success of any project depends on good planning. There is always the temptation to get out the tools and start making sawdust, but this often leads to mistakes that have to be undone later. The job will go more smoothly and the headaches will be fewer if you take the time to plot your course in advance by drawing plans and checking the requirements of local building codes. This chapter will help you get started.

Once your planning is done, you can begin to select materials. Wood will likely be chief among them. This chapter explains how wood is processed, how it behaves, and how to pick the right type for the job. You'll also be introduced to the many wood by-products and nonwood building materials currently available. Fasteners will have to be selected as well, and today that means more than just nails and screws. A section in this chapter discusses the many new options in fasteners and connectors.

Framing a roof is perhaps the most difficult, yet satisfying, basic carpentry skill. In some cultures the carpenters attach a flag, tree bough, or other traditional symbol to the highest point of the roof when they have completed framing it—and mark the occasion with a well-deserved celebration. As with any carpentry project, such success depends on plans that are made long before the first nail is driven.

PLANNING

The planning process involves evaluating your needs and then translating them into a comprehensive plan. This can be as simple as counting your books, measuring their overall dimensions, and drawing a plan for a bookshelf. For a major project, it's more extensive. You may need to enlist the help of design professionals.

Finding Ideas

If you have a general idea of a project you'd like to build but aren't quite sure what it should look like, you will first have to settle on a design. Spend some time on this phase, explore the possibilities, and allow the concept to incubate and mature before forging ahead. Poke through home-improvement books and magazines in local libraries and bookstores. Try to familiarize yourself with the different ways other people have solved problems similar to your own.

Working Drawings

Always try out ideas on paper before building anything. Even if you are working from a ready-made plan, chances are you will have to modify it to meet your particular needs. Drawing up the plan beforehand will help you visualize the construction process and anticipate any problems that might be encountered.

Make your drawings to scale on graph paper, with each square representing a certain number of inches or feet. Floor plans and elevations of buildings are usually drawn at a scale of ¼ inch to the foot. Construction details for foundations and framing connections are drawn at a larger scale, from ¾ inch to 1½ inches to the foot, depending on the level of detail required. Small woodworking projects are usually easier to draw full size.

Building Permits

If you are planning a substantial building project, you will need to take out a building permit. Any work that involves structural modifications to a house or changes to electrical, plumbing, or mechanical systems usually requires a permit—in fact, some municipalities require a permit for any job costing more than $100. Resist the temptation to go ahead without a permit. If you're discovered, the building inspector may require you to tear out all or part of what you've built and obtain a permit before you do it over again.

Make your first trip or a phone call to the building department before drawing the plans. Be sure to ask the following questions.

•What drawings are required? At the very least, you will need a site plan, a floor plan, and elevations. The site plan shows the property lines, the location of the house on the property, and the location of the planned improvements. A floor plan shows the location of walls, door and window openings, and plumbing and electrical fixtures. Elevations show side views of the outside of the building. In addition, you may need to provide details of foundations, framing connections, and major structural elements.

•What restrictions may apply to the project? Most jurisdictions have limitations on setbacks from property lines, overall building height, and lot coverage. In addition, some areas require new construction to meet specific architectural design criteria.

•What do permits cost? When you submit your plans, you must pay a fee based on the anticipated value of the improvements.

Budget and Schedule

Many home-improvement projects die an early death because the homeowners run out of time or money before they can finish. Don't rely on guesswork. Put together a budget and a schedule that are as realistic as possible.

Start with the materials list. Some lumberyards will figure the materials from your plans if the order is large enough. If you make your own list, try to think of everything you will need, from lumber to nails to hardware. Take your lists to potential suppliers for pricing.

Estimating labor is not so easy, especially if you're short on experience. The secret is to break the work down into manageable tasks and look at each one individually. Compile a list of everything you'll have to do, in as much detail as you can. Imagine yourself going through each phase of construction, one step at a time, and make an educated guess about the number of hours each step will take. Add up the total, and allow some extra time for setup, cleanup, problem solving and head scratching—and the inevitable run back to the lumberyard.

You can also arrive at a rough time estimate by applying this rule of thumb: On most construction projects, the cost of labor is approximately equal to the cost of materials. If you are a beginner, assume you will be working for minimum wage (pay yourself a little more if you have some carpentry experience). Divide the cost of materials by your hourly rate to figure the number of hours you can expect to spend with your tool belt on. If the result indicates that you will be giving up more of your spare time than you're willing to sacrifice, consider hiring professional help. You may want to subcontract any concrete work needed, for instance, and do the carpentry yourself.

Once you have a labor estimate, you can prepare a construction schedule. Making a schedule is helpful even for small projects because it provides clear-cut goals for each day's work and helps you avoid getting bogged down in small details. Remember, the object is to get from the beginning to the end of the project in the most direct way possible.

MATERIALS

Inevitably, material selection starts with wood. Frank Lloyd Wright summed up what so many feel. "Wood is universally beautiful to man," he said. "It is the most humanly intimate of all materials. Man loves his association with it, likes to feel it under his hand, sympathetic to his touch and his eye."

Wood is just about the nearest thing there is to the ideal building material. Pound for pound, it's nearly as strong as steel. It's easy to cut, shape, and fasten. It is warm to the touch—some kinds of wood even smell good. Its variations of color and grain are endlessly fascinating. All carpenters should be familiar with the properties of wood, what characteristics to look for when selecting lumber, and which products are appropriate for a given use.

From Trees to Lumber

It may be only a few hours from the moment a tree is felled until it arrives at the lumber mill for sawing. In less than five minutes, that tree can be converted into lumber ready for market.

Logs are first stripped of their bark, then are sent to the "head rig" to be sawed. The sawyer evaluates the log for quality and decides how it should be cut for the best yield. Then the log goes through a series of saws where it is sliced into slabs, rip-cut to width, and trimmed to length. Trimmings are fed into a chipper to be converted into raw material for paper, particle-board, and other wood prod-

ucts. In some modern mills, computers can analyze a log, plan the best cutting strategy, and set up the saws to make the cuts. The rough-sawed lumber is then smoothed, graded, and either stacked for shipment or set aside for kiln-drying and further processing.

The best cuts of wood come from the outside of old trees, near the bark. Knots are less numerous there, the grain rings closer together, and the wood more finely textured. These boards are graded for appearance and are cut into finish lumber and select lumber used for moldings and cabinet making.

Toward the center of the tree, where all branches originate, knots are more numerous. This part of the log yields the beams, timbers, and dimension lumber used for wood framing. This structural lumber is graded for strength rather than appearance.

Sawing Styles

How a log is cut affects the properties of the lumber produced. There are two basic methods of cutting wood: plain sawing and quarter sawing.

Plain sawing is the least wasteful and fastest way to saw a log. Most construction-grade lumber is produced in

this manner. Plain-sawed lumber is cut roughly parallel to the annual growth rings of the tree, producing boards with a grain pattern of wavy lines, V shapes, and ovals, with circular knots. Although this method yields strong lumber, plain-sawed boards can be prone to cupping and warping.

Quartersawed lumber, also called vertical-grain lumber, is cut perpendicular to the growth rings, resulting in boards with a grain pattern of parallel lines. As long as it is knot-free, quartersawed lumber is stronger and more stable than plain-sawed lumber, making it desirable for finish

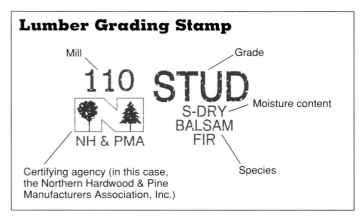

Lumber Grading Stamp

The boards in most lumber yards are sorted and stacked according to species, grade, and dimension. The numbers on the ends of these boards indicate length.

Wood-Cutting Methods

Plain-Sawed Lumber

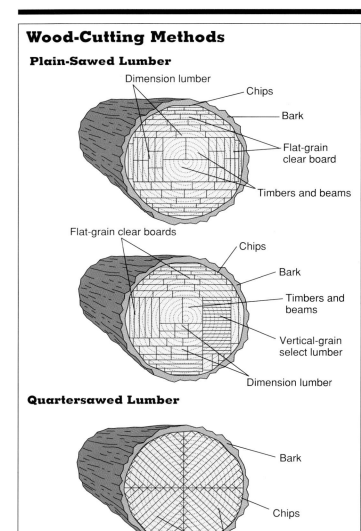

After debarking, a log is cut in either of two basic ways. The most common is called plain sawing. This process results in less waste, and therefore the wood is less expensive. The other process, called quarter sawing, results in lumber that is less likely to warp and is more beautifully grained in most species. There are further variations on these basic cutting methods, depending on the species, size of log, and kind of lumber ordered.

work. However, the knots in quartersawed lumber show up as long V-shaped spikes that may extend across the width of a board. A spike knot seriously weakens a board. Quarter sawing requires the kind of clear, defect-free logs that come mainly from old-growth forests. Because such forests are increasingly rare, and because so much of the log winds up as sawdust and waste wood, quartersawed lumber is fairly expensive.

Veneers are very thin sheets of wood that can be laminated to each other or to another substrate to produce panels. The resulting panels are strong and stable. Almost all veneers produced in the United States are made into plywood, although sheets of fine-wood veneers are available for furniture and cabinet making. Veneer is usually made by rotating a log against a fixed knife that peels off a continuous layer of wood the way paper toweling unspools from a roll. The grain pattern looks like a plain-sawed board that has been stretched out unnaturally wide. All structural plywood and the lower grades of finish plywood and paneling are made from rotary-cut veneers. For higher grades of finish plywood, the veneers are sliced off a fixed block of wood, resulting in natural-looking grain patterns that resemble plain-sawed or quartersawed lumber.

Wood Movement

Wood moves. It's an inescapable fact of nature, directly related to the amount of moisture in the wood. Freshly cut wood shrinks until it is dry and afterward expands and contracts with changes in humidity. To complicate things even more, wood changes its size significantly in width (across the grain), but hardly at all in length (along the grain).

In order to control shrinkage, the freshly cut, or green, lumber is either air-dried or kiln-dried. Initial air-drying brings the moisture content of the lumber down to about 20 percent when most of the free moisture in the pores of the wood has evaporated. Air-drying makes the wood stronger and only slightly shrinks it.

When the moisture content of lumber is taken below 20 percent, water trapped in the cell walls of the wood fibers begins to evaporate, the cells start to shrivel, and the wood shrinks. If the wood is being air-dried, this process can take a long time—about a year for each inch of thickness. In order to speed things up, mills dry lumber in kilns, where temperature and humidity are carefully controlled to prevent warping and cracking during drying. Kiln-dried lumber is much more stable than green lumber. It is also more expensive and therefore is usually reserved for high-quality cuts of wood used for furniture, cabinets, millwork, and floors.

In practical terms, the inherent tendency of wood to hold and relinquish moisture means that the header beams over doors and windows can shrink, cracking the wallboard in the corners. Doors and windows can get stuck in their frames. Plywood sheathing can swell and buckle. Board paneling and siding can shrink, causing unsightly gaps. Careful selection of lumber, storing lumber in a weatherproof location, and protecting bare lumber during the process of construction will keep moisture from being sealed into the job. In spite of these safeguards, permanent wood protection is often required.

Protecting Wood

Wood has two enemies: fungus and bugs. Several types of fungus attack wood. Fungal infestations are often referred to as dry rot. The name is something of a misnomer, because dry-rot fungi can survive only under damp conditions, when the moisture content of wood is over 20 percent. Therefore, it is important to keep wood buildings dry by providing air circulation around structural members (particularly under floors), sealing against the intrusion of water, and maintaining proper clearances from moist earth.

Insects can also cause serious damage to wood buildings. Of all the wood-eating bugs, subterranean termites (genus *Reticulitermes* sp.) do the most damage. These termites build their nests underground and travel back and forth to their feeding grounds through mud tunnels. You can often spot the source of an infestation by looking for tunnels along foundation walls and in crawl spaces. Termites are attracted to moist wood, so keeping wood dry is your best first line of defense. Treating serious infestations is a job best left to experienced professionals.

Because most dry-rot and termite problems start close to the ground, use pressure-treated wood, which is immune to both fungi and termites, in locations where wood is in contact with foundations, slabs, or bare earth.

Decay-Resistant Wood Species

Some types of trees have an inherent preservative (usually tannic acid) that protects them from decay. Only the darker heartwood, from the center of the tree, contains this preserving agent; the lighter colored sapwood, surrounding the heartwood, is not rot resistant. Redwood, cypress, and some (but not all) species of cedar are the most common examples of decay-resistant lumber.

Because of their beautiful coloration and excellent weathering properties, these woods are prized for exterior woodwork. They are also more expensive, and their use is usually restricted to siding, railings, trim, and other applications where appearance and durability are most important.

Pressure-Treated Wood

Wood species that would normally be subject to decay and insect attack can be treated by a process of injecting preservatives under pressure. These preservatives bind chemically to the wood tissue and won't leach out, making the wood nontoxic under normal use. Pressure-treated lumber is far superior to wood that has been merely sprayed or dipped in a preservative. In many cases, it will outlast naturally durable species. Pressure-treated lumber is necessary when wood will be subjected to ground contact or will be buried.

The most common types of chemicals used for pressure-treating lumber are waterborne salts such as chromated copper arsenate (CCA) and chemonite, or ammoniacal copper arsenate (ACA). These preservatives are appropriate where the wood will be used close to plants. They can be used safely around the home except for surfaces that will be in direct contact with food or serving utensils. Other preservatives used are pentachlorophenol and creosote, but they are so toxic that their use has been banned in many parts of the country. Lumber treated with these preservatives cannot be painted or stained as easily as lumber treated with CCA or ACA. Pentachlorophenol applied with liquid petroleum gas (LPG) is an exception.

Pressure-treated lumber has a green or beige tint and does not darken if left to weather. Depending on the amount of preservative used, pressure-treated wood is rated for ground contact (LP-22 rating, for direct burial uses, such as fence posts), or above-grade use (LP-2 rating, for sills, bottom plates of walls, and framing members within 6 inches of the ground). Sometimes pressure-treated wood is incised or punctured on the surface to facilitate the penetration of chemicals. For projects where a smooth, unblemished appearance is critical, ask for pressure-treated lumber without incisement. It's also worth the extra cost to buy lumber that is kiln-dried after treatment (KDAT) to avoid excessive warping in exposed outdoor applications.

Wear goggles and a dust mask when cutting pressure-treated lumber and gloves when handling it—especially if the lumber is damp. Do not burn scraps; dispose of them in an approved landfill. Coat the ends of cut boards with an approved preservative.

Lumber Grades

Virtually all the lumber produced in the United States is subject to grading systems that evaluate wood products for strength and appearance. Graders base their judgments on such factors as wood species, number and type of defects, grain patterns, and surface appearance. There are many grading associations across the country, each with its own set of rules, so you can't assume that a given lumber grade always means the same thing. With this in mind, consider the information below a *general* guide to lumber grading as practiced by most associations. Check with a local lumber supplier to verify that what you are buying is suitable for your needs.

For grading purposes, there are two broad categories of softwood lumber: structural lumber, which is graded primarily for strength, and finish lumber, which is graded for appearance.

Structural Lumber

This is the lumber used to construct wood-framed houses. Pieces of wood 2 to 4 inches thick are known as dimension lumber. The term *timber* includes anything 5 inches or thicker.

Smaller sizes of dimension lumber (2×4 and 4×4) are graded differently than the larger sizes. Utility grade is

Characteristics of Wood

These charts give detailed information on softwoods and hardwoods. The softwoods are used in rough carpentry; the hardwoods are used more often in fine woodworking. For the construction process described in this book, you will be concerned mainly with softwoods. You'll find the chart on hardwoods helpful for your other woodworking projects.

Softwood

Type	Sources	Uses	Characteristics
Cedar (Red)	East of Colorado and north of Florida	Mothproof chests, lining for linen closets, sills.	Very light, soft, weak, brittle. Natural color. Generally knotty, beautiful when finished in natural color, easily worked.
Cedar (Western Red, White)	Pacific Coast, Northwest, Lake States and Northeastern United States	Paneling, fence posts, siding, decks. Red cedar used for chests and closet lining.	Fine-grained, soft. Red: reddish brown with white sapwood. White: light with light brown heartwood. Brittle, light-weight; easily worked; low shrinkage; high resistance to decay. Strong, aromatic.
Cypress	Southeastern coast of the United States	Interior wall paneling and exterior construction, i.e., posts, fences, cooperage, docks, bridges, greenhouses, water towers, tanks, boats, river pilings.	Lightweight, soft, not strong, easy to work. Coarse texture. Durable against water decay. Light brown to nearly black.
Fir (Douglas)	Washington, Oregon, California	Construction flooring, doors, plywood, low-priced interior and exterior trim.	Light tan; moderately hard; close-grained. Most plentiful wood in the United States; used mostly for buildings and structural purposes; strong, moderately heavy.
Fir (White)	Idaho, California	Small home construction.	Soft; close, straight-grained; white with reddish tinge. Low strength; nonresinous; easily worked, low decay resistance.
Hemlock	Pacific Coast, western states	Construction lumber; pulpwood; containers; plywood core stock.	Lightweight; moderately hard; light reddish brown with a slight purple cast.
Larch	East Washington, Idaho, Oregon, Montana	Framing, shelving, fencing, shop projects, furniture.	Moderately strong and hard. Glossy russet-colored hardwood with straw-colored sapwood. If this wood is preservative treated, it can be used for decking.
Pine (Lodgepole)	West Coast from the Yukon to Mexico	Framing, shelving, fencing boards, small furniture, shop projects.	Hard, stiff, straight-grained. Mills smoothly, works easily, glues well, resists splintering, and holds nails well. Light brown heartwood, tinged with red with white sapwood.
Pine (Southern)	Southeastern United States	Floors, trusses, laminated beams, furniture frames, shelving.	Strongest of all softwoods with a pale yellow sapwood and reddish heartwood.
Pine (Sugar)	Western Oregon's Cascade mountains, Sierra Nevada of California	Shingles, interior finish, foundry patterns, models for metal castings, sash and door construction, quality millwork.	Lightweight, uniform texture, soft. Heartwood is light brown with tiny resin canals that appear as brown flecks. The sapwood is creamy white. Straight-grained and warp resistant.
Pine (White, Ponderosa)	United States	Solid construction in inexpensive furniture, sash, frames, knotty paneling.	Soft; pale yellow to white in color; fine-grained; darkens with age. Uniform texture and straight grain; lightweight, low strength, easily worked; has moderately small shrinkage, polishes well; warps or swells little.
Redwood	West Coast	Sash, doors, frames, siding, interior and exterior finish, paneling, decks.	Heartwood is cherry to dark brown; sapwood is almost white. Close, straight grain. Moderately lightweight, moderately strong; great resistance to decay; low shrinkage, easy to work, stays in place well; holds paint well.
Spruce	Various parts of the United States and Canada	Indoor work only. Pulpwood; light construction and carpentry work.	Soft, lightweight, pale with straight unpronounced grain and even texture.

Hardwood

Type	Sources	Uses	Characteristics
Ash (White)	United States	Upholstered furniture frames, interior trim.	Hard; prominent, coarse grain; light brown. Strong, straight grain; stiff, shock resistant; moderate weight; retains shape, wears well; easily worked.
Beech	Eastern United States	Flooring, chairs, drawer interiors.	Hard; fine grain; color varies from pale brown to deep reddish brown. Heavy, strong; has uniform texture; resists abrasion and shock; medium luster.
Birch (Yellow, Sweet)	Eastern and Northeastern United States and Lake States	Cabinet wood, flooring, plywood paneling, exposed parts and frames of furniture.	Hard, fine grain; light tan to reddish brown. Yellow is most abundant and important commercially; white sapwood and reddish brown heartwood. Heavy, stiff, strong; good shock resistance, uniform texture; takes natural finish well; satiny in appearance.
Cherry (Black, Wild)	Eastern and Northern United States	Paneling, furniture.	Moderately hard; light to dark reddish brown, fine grain; darkens with age. Strong, stiff, heavy; high resistance to shock and denting; not easily worked; high luster.
Gum (Red, Sap)	South	Plywood, interior trim, posts, stretchers, frames, supports—frequently used in combination with other woods.	Heartwood (red gum) is light to deep reddish brown; sapwood (sapgum) nearly white; moderately hard, fine grain. Moderately heavy, strong, uniform texture; takes finish well; frequently finished in imitation of other woods. In early 1900s, it was the most frequently used furniture wood in the U.S.
Hickory	Arkansas, Ohio, Tennessee, Kentucky	Tool handles, wagon stock, baskets, wagon spokes, pallets, ladders, athletic goods.	Very heavy, hard, stronger and tougher than other native woods; sapwood and heartwood of same weight. Difficult to work, subject to decay and insect attack.
Maple (Sugar, Black)	Great Lakes, Northeast, Appalachians	Interior trim, furniture, floors in homes, dance halls, bowling alleys.	One of the hardest woods in U.S.; heartwood is reddish brown; sapwood is white; usually fine, straight-grained; sometimes curly, wavy, or bird's-eye grain occurs. Strong, stiff; good shock resistance; great resistance to abrasive wear, one of the most substantial cabinet woods; curly maple prized for fiddlebacks.
Oak (Red, White)	Eastern United States, mainly Mississippi Valley and South	Flooring, interior trim, furniture, plywood for cabinetwork, paneling.	Hard, pronounced open grain; rich golden color to light reddish brown. Moderately heavy, stiff, strong, resilient, tough; comparatively easy to work with tools; takes many finishes.
Poplar (Yellow)	Eastern United States	Interior trim, siding, furniture, panels, plywood cores.	Sapwood is white; heartwood is yellowish brown tinged with green; soft; straight, fine grain. Lightweight, moderately weak, does not split readily when nailed; easily worked, easy to glue; stays in place well, holds paint and enamel well; finishes smoothly.
Sycamore	Eastern half of United States	Interior trim, fancy paneling, furniture.	Light to reddish brown; hard, close, interlocked grain. Moderately heavy, strong; rays are conspicuous when quartersawed; seasoning without warping is difficult.
Walnut (Black)	Central United States	Furniture, paneling, cabinetwork interior.	Light to dark chocolate brown; hard; moderately prominent, straight grain; sapwood is nearly white. Strong; resists shock and denting, easily worked; takes stain and finishes exceedingly well; heavy, stiff, is stable in use; one of the most beautiful native woods; has luminous finish.

the lowest grade suitable for construction. Its use is restricted to nonstructural building components such as interior partitions, blocking, and furring strips. The next rung up the quality ladder is standard grade, the most common grade used for light framing. Construction-grade lumber is stronger yet; it is straighter and contains fewer and smaller defects than the other grades. A fourth grade of small-size dimension lumber is called stud grade. Intended for wall framing, it has a strength rating that usually falls somewhere between standard and construction grade. Most stud-grade lumber comes precut to the exact length required for common sizes of stud walls.

In the larger sizes of dimension lumber and timbers (2×6 and up), there can be as many as 10 different grades to choose from. For residential construction, three grades are all you will ordinarily need. Grade 2 lumber is all-purpose lumber suitable for most structural purposes. It may contain heart center (the middle of the tree), which can cause warping, cracks, and twisting, but in applications that are concealed by the finished structure, this is not usually a serious problem. Grade 1 lumber is generally free of heart center. It also has straighter grain and smaller knots, and it's used where appearance or extra strength are important. At the top of the structural lumber heap is select structural grade. The strongest, stiffest, and most stable of all grades, it is used for heavily loaded beams and

Nominal and Actual Dimensions of Lumber

Nominal Size	Actual Size*
1 × 2"	¾ × 1½"
1 × 3"	¾ × 2½"
1 × 4"	¾ × 3½"
1 × 6"	¾ × 5½"
1 × 8"	¾ × 7¼"
1 × 10"	¾ × 9¼"
1 × 12"	¾ × 11¼"
5⁄4 × 6"	1¼ × 5½"
2 × 2"	1½ × 1½"
2 × 3"	1½ × 2½"
2 × 4"	1½ × 3½"
2 × 6"	1½ × 5½"
2 × 8"	1½ × 7¼"
2 × 10"	1½ × 9¼"
2 × 12"	1½ × 11¼"
4 × 4"	3½ × 3½"
4 × 6"	3½ × 5½"
4 × 8"	3½ × 7¼"
4 × 10"	3½ × 9¼"
4 × 12"	3½ × 11¼"
6 × 6"	5½ × 5½"
6 × 8"	5½ × 7¼"
8 × 8"	7¼ × 7¼"

*Dimensions may vary. Always measure.

horizontal members that have to span long distances. It is also the most expensive grade, but is worth the extra money if you need good lumber for exposed structural framing.

Finish Lumber

Grading systems for finish lumber are not standardized and vary according to wood species and the various grading associations involved. Some use a letter system, with grade A being the best grade and grade D the lowest; others use numbers, beginning with grade 1. The part of the tree the wood was cut from also

Problem Boards

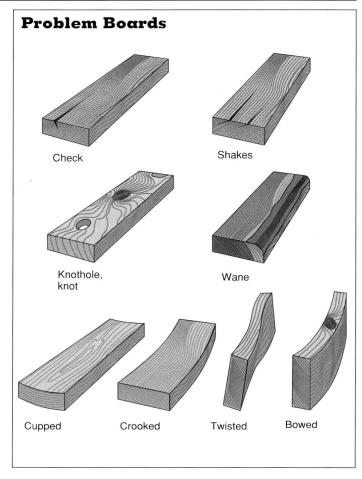

Check

Shakes

Knothole, knot

Wane

Cupped

Crooked

Twisted

Bowed

comes into play, because the heartwood of some species is more valuable than the sapwood. Quarter sawing, moisture content, and other factors also affect how finish lumber is graded.

Remember that finish lumber is graded for appearance. Select your own finish lumber whenever possible. If it looks good to you, and the price is right, it will probably do the job.

Selecting Lumber

Wood comes from living trees and is bound to contain some defects. A perfect board is rare

indeed. Nevertheless, if you understand the nature of common wood defects and how they affect wood performance, you can make informed decisions when the time comes to pick out the lumber for your project.

• Knots are not necessarily defects. In fact, a tight knot in the center of a board may actually make it stronger, because the wood tissue around the knot is denser than the rest of the board. A board with an edge knot, on the other hand, will eventually develop a kink at the knot, even if the board looks straight when you buy it. Loose knots—the kind that fall out and leave an open hole—are acceptable as long as they are small, but large ones should be considered serious defects. Spike knots, sometimes found in quarter-sawed lumber, cause problems if their diameter is more than one fourth of the thickness of the board. Set aside lumber with unacceptable knots and cut for blocking, cripple studs, and other short pieces.

• Cupped structural lumber is usually not a problem. All plain-sawed lumber cups to some extent, especially if it is milled while still green.

Cupping can be a problem with finish lumber, though, especially if it's used outside, where it will be exposed to the weather. Your best insurance against cupping is to buy lumber that has been kiln-dried before surfacing. An even better solution, if your budget permits, is to buy quarter-sawed lumber—it's least likely to have cupping problems.

• Twisted lumber causes the most problems when used for header beams over door and window openings. No matter how you install a twisted header, one or more corners will stick out and cause lumps in the wall finishes. For studs, joists, and rafters, twist can usually be eliminated by blocking and strategic nailing. There's not much you can do about twisted beams and timbers, though, so avoid them unless you can use them where they won't show.

• Bowed lumber is nothing to worry about unless the bow is really pronounced. Bowed studs can be straightened by nailing a piece of scrap to adjacent studs until the wall finishes are in place, or by

adding extra blocking. Since finish lumber is usually nailed to the framing, nails will hold bowed pieces flat.

• Crooked lumber is not uncommon and is hard to avoid. You just have to make allowances for it when you use it. You can spot even a very slight crook by picking up one end of a board and sighting down it with one eye. The convex edge of the board is called the crown side. If you're framing a wall, arrange all the studs so the crowns face the outside of the building. For roofs and floors, place the crowns up so that they will straighten out when the building is loaded. Badly crooked lumber should be rejected or cut into shorter pieces.

• Checks are splits that run perpendicular to the growth rings of a piece of lumber. These cracks that result from shrinkage usually show up near the ends of boards, where the wood dries out fastest. Lumber that contains boxed heart, or the center of the tree, checks more than other cuts, and unseasoned lumber checks more than kiln-dried wood. Still, checking is more of a cosmetic than structural defect, so don't hesitate to use checked lumber if appearance is not the primary consideration.

• Shakes, which are splits that follow the grain rings, are more serious. They are caused by extreme bending stresses in the tree before it is cut, usually from wind or snow loads. Do not use lumber with shakes that extend more than halfway through the board.

• Wane is the natural outer surface of the tree, just under the bark, that is found in some

lumber. Because boards with wanes at the edges are cut from the outside of the log, they are likely to be of high quality, except for the bad edge. As long as there is enough of a flat surface on this edge to hold a nail, lumber with wane is acceptable for most uses.

Storing Lumber

Even if you are able to purchase the best-quality lumber, it's important to store it properly until it can be used. Lumber that's piled up willy-nilly will turn into worthless wood spaghetti in very short order. Stack lumber off the ground on pieces of scrap wood, called stickers, with all pieces butted tightly together. Cover the stack with plywood or a weighted tarp.

If you won't be using the lumber for a period of weeks or months, it's a good idea to store the whole pile with stickers. This promotes uniform seasoning by allowing air to circulate around every piece in the stack. Place strips of lath, spaced about 4 feet apart, between each layer of lumber. Make sure each sticker is directly above the one below it, or the weight of the pile will put a permanent bow in the boards at the bottom.

Manufactured Wood Products

Manufactured wood products are produced by taking wood apart and reassembling it into new forms with high-strength

adhesives. Manufacturers of these products can use marginal lumber and milling waste to make building materials that in many ways are superior to their solid-wood counterparts.

You are probably familiar with some of these materials—plywood, particleboard, and hardboard have been around for many years. Others, such as oriented-strand board (OSB) and laminated-veneer lumber (LVL), are less familiar but are becoming more common.

Some manufactured wood products are bonded with urea-formaldehyde or phenol-formaldehyde glues. These resins emit gases for a significant period of time. People who are sensitive to urea-formaldehyde—used in particleboard and some types of medium-density fiberboard (MDF)—should choose a product made with less noxious phenolformaldehyde, and test their reaction to it before building an addition or a home. When buying particleboard or MDF, check for the mark HUD 24 CFR PART 3280. This indicates that the product complies with federal standards on emissions of formaldehyde gas.

Sheet Products

Plywood

This familiar material has long been widely used in the building trades. Layers of veneers are glued up into sheets, with the grain direction of the veneers alternating between layers. These sheets are stable, split resistant, and big enough to cover large areas quickly.

Plywood Veneer Grades

This chart with its description of the six different levels of veneer grades on plywood, listed in descending order of quality, was made available by the American Plywood Association.

N	Smooth-surface "natural finish" veneer. Made of select-grade wood that is either all heartwood or all sapwood. Free of open defects. No more than 6 repairs permitted in each 4 × 8 panel; each must be made of wood, must be parallel to the grain, and must be matched to the grain and color of the panel. This is top-quality grade.
A	Smooth and paintable with no more than 18 neatly made repairs; each must be made parallel to the grain. Can be used for natural finish in applications that are not too demanding.
B	Solid surface with circular repair plugs, shims, and tight knots up to 1 inch across the grain allowed. Some minor splits permitted.
C plugged	Improved C veneer. Splits limited to ⅛-inch width with knotholes and borer (insect) holes limited to ¼–½ inch. Some synthetic repair and broken grain permitted.
C	Tight knots to 1½ inches allowed. Knotholes permitted up to 1 inch across the grain and some to 1½ inches if the total width of the knots and knotholes is within specific limits. Permissible to have synthetic or wood repairs, discoloration, and sanding defects if they do not impair the strength. Limited splits are allowed. Stitching—the process of sewing random-sized pieces of plywood together—is permitted.
D	Knots and knotholes across the grain and up to 2½ inches wide are allowed. Within specified limits they can be up to 3 inches wide. Stitching is permitted as is a limited number of splits. This level is limited to interior grades of plywood.

Plywood manufacturers have developed dozens of different grades and types of plywood for specific uses, but there are basically two general categories: structural plywood and finish plywood.

Structural plywood is usually made of softwood species and always carries a grade stamp that gives the structural grade, allowable rafter and floor-joist spacing, and other important information. It's most often used for sheathing floors, roofs, and walls.

Finish plywood has face veneers of hardwood or softwood, and is designed to be used for finished products such as cabinets, wall paneling, and door skins.

Oriented-Strand Board and Waferboard

These products are made from large wood chips fused together with phenolformaldehyde adhesives. Oriented-strand board (OSB) consists of layers of chips laid so the fibers in one layer run at more or less right angles to those in the next, similar to the way plywood is made. The chips used in waferboard are randomly oriented in the sheet. OSB is stronger than waferboard for a given thickness, so it is the type most often found on building sites.

OSB panels are structurally as strong as plywood and are usually used for roof sheathing, wall sheathing, and subfloors. The panels have a smooth side and a textured side. The textured side should be installed up if you use OSB for roof sheathing, because the smooth face can be dangerously slippery.

Because OSB expands more than plywood with changes in moisture, the installed panels should be spaced ⅛ inch apart on the edges and ¼ inch at the ends. OSB is not as good as plywood at holding nails, especially near the edges. Therefore, it's important to maintain the proper distance from the edges of the sheet when nailing. OSB also tends to swell around the edges when it gets wet and therefore may not be the best choice for underlayment beneath thin floor coverings.

Particleboard and Medium-Density Fiberboard

These two products are made from wood chips, sawdust, and small fibers, bonded with urea-formaldehyde or phenol-formaldehyde adhesives, and compressed into sheets under heat and pressure.

Particleboard, composed of small chips and sawdust, is an inexpensive alternative to plywood for many applications. It is dense, heavy, and impact resistant, making it a good choice for underlayment beneath sheet flooring, for concealed portions of cabinets, and as the substrate for plastic-laminate countertops.

Medium-density fiberboard (MDF) is made from individual wood fibers bonded with phenolformaldehyde adhesives. It is very finely textured and even denser than particleboard. MDF has excellent machining properties and is used to make paint-grade moldings, door panels, and furniture parts.

The edges of particleboard and MDF sheets are poor at holding nails and screws; you must use specialized fasteners or pieces of solid wood when joining pieces of particleboard or MDF.

One useful particleboard variant is known as melamine board. This is a particleboard sheet that has been coated with a layer of tough, lightly textured plastic. The primary advantage of this material is that it can be made into cabinets that don't require further finishing. Another plus is that the melamine coating seals the particleboard surface, virtually eliminating any off-gassing of formaldehyde vapors. Exposed edges of melamine boards are usually covered with wood or plastic veneers or with strips of solid wood. White, cream, and gray are stock colors, but melamine can be ordered in dozens of other colors.

Composition Panels

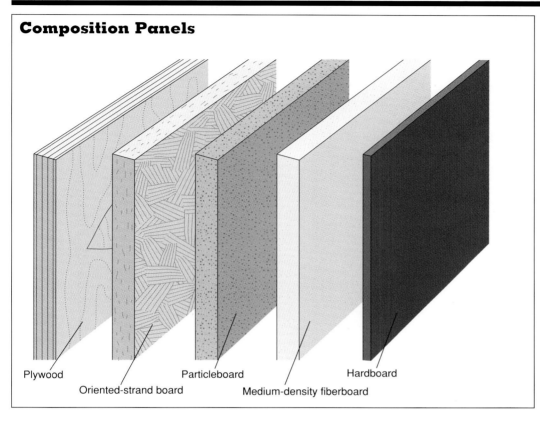

Plywood

Oriented-strand board

Particleboard

Medium-density fiberboard

Hardboard

Manufactured Beams and Framing

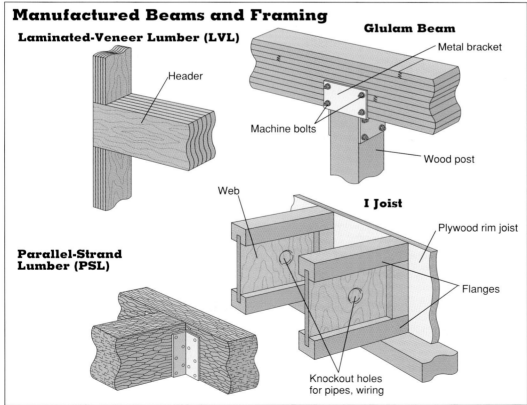

Laminated-Veneer Lumber (LVL)

Header

Glulam Beam

Metal bracket

Machine bolts

Wood post

Parallel-Strand Lumber (PSL)

Web

I Joist

Plywood rim joist

Flanges

Knockout holes for pipes, wiring

Hardboard

Hardboard, which is actually a high-density fiberboard, is used for cabinet backs, drawer bottoms, siding, perforated board, and doors. It comes tempered and untempered. The tempered variety is water resistant and is used to make siding.

Unlike solid wood, hardboard siding moves in length as well as width, so special installation procedures must be followed to allow for expansion and contraction. Follow the manufacturer's instructions carefully to protect the warranty when installing hardboard siding.

Manufactured Beams and Framing Lumber

Wood I Joists

I joists are made of a plywood or OSB web glued to two solid-wood or LVL flanges in the shape of an I beam. They are primarily used for floor joists, although they can also be used for rafters, ceiling joists, and headers.

Wood I joists offer many advantages over solid-wood framing materials. They are lighter, straighter, and more dimensionally stable than comparable sawed lumber. Floors framed with I joists are flatter than conventionally framed floors and are not subject to shrinkage that can cause settling and annoying floor squeaks. I joists are capable of very long spans—up to 25 feet or more—and can eliminate the need for intermediate beams and bearing walls.

Framing With Steel

As the available supply of quality timber dwindles and prices rise, builders have been exploring alternatives to traditional wood framing. One such option that has seen rapid growth in recent years is the steel-stud framed building.

Steel framing has several advantages over wood. Steel studs are lightweight and strong. They won't burn (although they can lose their strength in a very hot fire). They won't shrink, warp, or rot, and termites hate them. Steel studs can be made from recycled scrap metal and can be recycled again once their useful life is over. And every house framed in steel saves the lives of several trees.

The basic structural components of steel-framed buildings are similar to their wood counterparts, with studs, joists, rafters, and roof trusses, but the techniques used to assemble them are different. Hammers and nails give way to electric screw guns and drill-point screws, and toothed saw blades are replaced by abrasive metal-cutting blades and tin snips.

Steel studs are made of C-shaped sheet-metal channels. They come with pre-punched holes for running electrical wiring and plumbing through the walls. Steel studs are available in several different sizes and weights, depending on load-bearing requirements.

Virtually all connections in steel framing are made with specially designed screws. For light-duty work, a sharp-pointed screw that punches through the sheet metal is used; for heavy-gauge studs, a drill-point screw, with a tip that looks just like a small drill bit, is the fastener of choice. Other specialized screws are used to fasten the wallboard, siding, floors, and trim to the framing. Given the great number of screws in a metal-stud job, the only practical way to drive them is with a power screw gun.

Although sizable load-bearing structures require specialized building techniques, nonbearing interior walls built of light-gauge studs are easy to construct with just a screw gun and metal snips (use a felt-tip pen for layout work). Metal studs can be a good choice for remodeling jobs, with the added bonus of not having to sweep up a pile of sawdust at the end of the day. If you are thinking of framing a project in steel, obtain a copy of the manufacturer's recommended framing practices and study it carefully. Consult an architect or engineer for any job with significant structural loads.

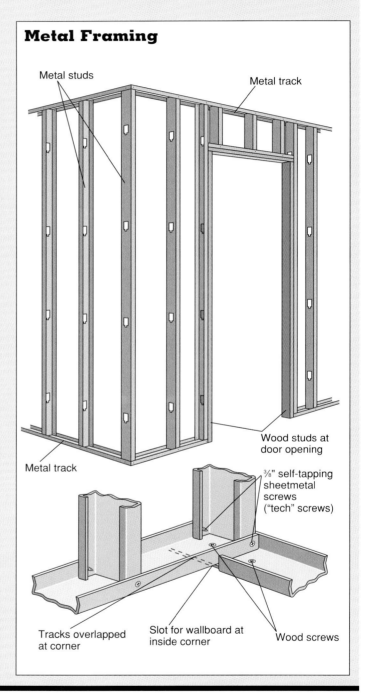

Metal Framing

Metal studs

Metal track

Wood studs at door opening

Metal track

⅜" self-tapping sheetmetal screws ("tech" screws)

Tracks overlapped at corner

Slot for wallboard at inside corner

Wood screws

Holes can be cut in the webs to route plumbing, wiring, and duct work through the floor. When properly placed, these holes can be nearly the full depth of the joist.

If you are thinking of using I joists, bear in mind that it's crucial to follow proper framing procedures. Most manufacturers provide detailed drawings of typical framing conditions and span charts for normal uses. The supplier will put together a complete materials package from your drawings and specify which framing details are appropriate for your installation.

Laminated-Veneer Lumber

This material, known as LVL, consists of layers of veneer glued up like plywood; unlike plywood, the grain direction of all the plies runs the same way. This results in lumber that is stronger for its size than any other wood product. LVL is manufactured in a standard thickness of 1¾ inches, with widths to match standard framing lumber dimensions.

LVL is typically used for beams, headers, and girders. Several pieces can be stacked side by side to build up heavy members of any dimension. The great strength of LVL makes it ideal for headers over wide openings, such as garage doors, or as flush girders in floor systems. Because the face veneers are not selected for appearance and are usually covered with glue stains, LVL is most often used in areas that will be concealed within the finished structure.

Glued-Laminated Timbers

These large beams, called glulams, are made by gluing up stacks of standard dimension lumber. Once assembled, the beams are surfaced and sanded, so they can be used to good effect in exposed locations.

Glulams are stronger than sawed timbers of the same size, can be much longer—up to 60 feet or more—and do not shrink or warp easily. Most glulams have a slight curve built into them to compensate for sag when loaded, so be sure that you install them with the edge marked "top" facing upward.

Connect glulams to other framing members with metal brackets and machine bolts. Be sure the brackets you use are sized for glulams, which are narrower than sawed timbers of similar size. Notching or drilling through a glulam may weaken it significantly, so ask for the manufacturer's specifications for allowable cutting and notching.

Parallel-Strand Lumber

Known as PSL, this lumber is made by fusing long wood chips together in such a way that all the fibers run in the same direction. The resulting material is strong, more stable than sawed lumber, and, like glulams, can be purchased in long lengths. PSL is manufactured in straight pieces—without the camber glulams have—and therefore may sag a little under heavy loads. Because they have a pleasing, uniform surface texture, PSL beams can be used in exposed locations where appearance counts.

Plastic Wood

The name sounds like heresy to purists, but this recently developed product has its place in home construction. Made of recycled plastic, sometimes combined with sawdust filler, plastic wood can be cut, shaped, and nailed like solid wood. It is not approved for structural uses. Plastic wood is best suited to applications that are subject to heavy weathering such as deck planking, garden borders, and signposts.

Plastic Lumber

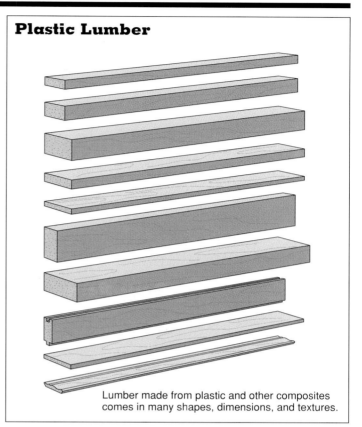

Lumber made from plastic and other composites comes in many shapes, dimensions, and textures.

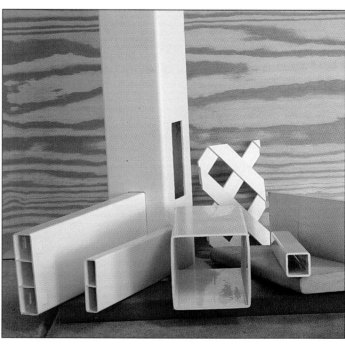

This sampling of alternatives to wood includes hollow vinyl "lumber" used for fences, solid vinyl lattice, and boards composed of wood fibers and recycled plastic.

FASTENERS

There are literally hundreds of different fasteners designed to connect building components. Each type has its own best use. The following guide lists the most common fasteners used in residential construction.

Nails

Each type of nail comes in various lengths, designated by the term *penny (d)*. This traditional English term originally indicated the cost of 100 nails of a given size. Nowadays, nails are sold by the pound.

Always use corrosion-resistant nails in locations that will be exposed to the weather. Galvanized (zinc-coated) box nails are the most common type. Hot-dipped galvanized nails (denoted HDG on the box) are more reliable than electrogalvanized (EG). Aluminum, stainless steel, and bronze nails also offer corrosion protection, although at a considerably higher price.

Basic Nails

For general framing work, use box, sinker, or common nails. Common nails are the heaviest type. They have the most holding power, but they're hard to drive and they tend to split the wood. Box nails, with their narrow shanks, cause less splitting but bend easily. Sinkers, which have a shank thicker than a box nail but thinner than a common nail, are the easiest to drive. Nails designated "cement-coated" or "vinyl-coated" have a thin coating of adhesive that increases their holding power.

Finishing nails have slender shanks and small, barrel-shaped heads that can be driven below the surface of the wood with a nail set. Use them for installing doors, windows, paneling, trim—anywhere you don't want the heads to show. Casing nails, which are heftier versions of finishing nails, provide more holding power. Brads are miniature finishing nails used for attaching thin, fragile pieces of wood.

Special-Purpose Nails

There are many other types of nails for specialized applications. Some are shown below.

• Roofing nails are galvanized nails with wide heads. They are used for building paper and composition roofing materials. Aluminum roofing nails are used for aluminum roofing and flashings (ordinary galvanized nails will corrode if they are used with aluminum).

Nails

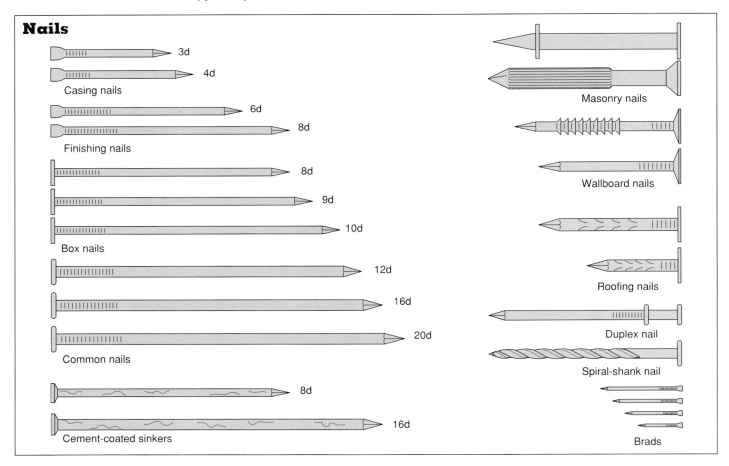

Casing nails — 3d, 4d

Finishing nails — 6d, 8d

Box nails — 8d, 9d, 10d

Common nails — 12d, 16d, 20d

Cement-coated sinkers — 8d, 16d

Masonry nails

Wallboard nails

Roofing nails

Duplex nail

Spiral-shank nail

Brads

Driving Nails

Hold the nail near the head

"Sight" your target by touching hammer to nail as you want to hit it

A sign of poor hammering

• Duplex nails have double heads for easy removal. Use them for scaffolding, concrete forms, and temporary bracing.

• Masonry nails, made of hardened steel, can be driven into concrete or masonry.

• Wallboard nails are designed to have enough holding power to keep them from popping out of a wall. One type has a chemically etched shank and a concave head. Another has a ring shank and a flat head.

• Spiral-shank and ring-shank nails have shanks designed for extra holding power. Used mainly for floors, they are also good for exterior siding and trim when wood movement would cause ordinary nails to work loose.

Tips for Choosing and Using Nails

Length

In general, choose a nail that is about three times longer than the thickness of the material to be fastened. For example, to install ¾-inch-thick trim, use a nail that is approximately 2¼ inches long. The exception to this rule is when you are nailing 2-by framing members together; here, the fastener of choice is the 16-penny (3½-inch) nail.

Driving Nails

Start nails with a light tap of a hammer. Hold the nail near the head, not down by the point. That way, if you miss, you're less likely to smash your fingers. Then move your hand out of the way and drive the nail home with smooth, deliberate hammer strokes, swinging from the shoulder, not the elbow. Keep your eyes focused on the head of the nail, which will help you guide the hammer to its mark.

Preventing "Hammer Tracks"

Even experienced carpenters miss a nail once in a while. This isn't a problem for rough framing, because your work will be covered up later. For finish work, however, dents from missed hammer blows are ugly—and embarrassing.

Here's a trick to keep from leaving "hammer tracks" on your work. Drill a hole near one end of a scrap of plywood or hardboard. Start the nail normally, then slip the hole over the head of the nail. Drive the nail as far as you can—if you miss, you will only dent the piece of scrap. Remove the scrap and finish off the nail with a few light hammer blows.

Preventing Splits

Sometimes wood splits when you drive a nail into it—especially near the ends of a board. To prevent this, keep your nails at least one nail length from the end of the piece whenever possible. If the problem persists, place the nail upside down on a hard surface and tap the point with a hammer. This produces a blunted point that will punch its way through the wood fibers instead of spreading them apart. An even better solution is to drill a pilot hole first. Use a drill bit that is approximately three fourths of the diameter of the nail shank.

Screws

The three broad categories of screws used in construction are wood screws, sheet-metal screws, and wallboard screws, but different sizes, heads, points, and materials create dozens of variations.

Wood Screws

Made of mild steel or brass, wood screws have a shank that is threaded for about two thirds of the screw's length. Most wood screws are available with round, oval, or flat heads for driving with Phillips or slotted screwdrivers.

Sheet-Metal Screws

These screws are made of hardened steel and are threaded all the way up to the head. Pan-head sheet-metal screws are commonly available with either slotted or Phillips heads. Some specialty types have a hex head for driving with a nut driver, and others have a drill-point tip that drills its own pilot hole.

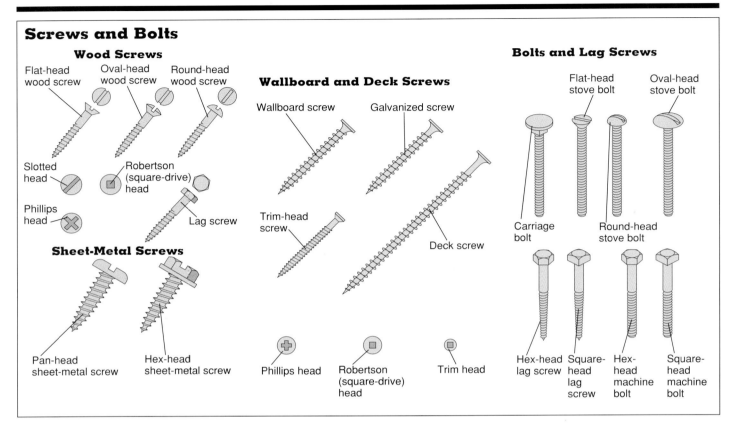

Screws and Bolts

Wood Screws

Flat-head wood screw
Oval-head wood screw
Round-head wood screw

Slotted head
Robertson (square-drive) head
Phillips head

Sheet-Metal Screws

Pan-head sheet-metal screw
Hex-head sheet-metal screw

Wallboard and Deck Screws

Wallboard screw
Galvanized screw
Trim-head screw
Deck screw
Lag screw

Phillips head
Robertson (square-drive) head
Trim head

Bolts and Lag Screws

Flat-head stove bolt
Oval-head stove bolt
Carriage bolt
Round-head stove bolt
Hex-head lag screw
Square-head lag screw
Hex-head machine bolt
Square-head machine bolt

Wallboard Screws

Known as zip screws, general-purpose screws, drive screws, or bugle-heads, these fasteners have hundreds of uses. Originally developed for fastening wallboard to framing, they have evolved into many new variations for a wide range of uses. These fasteners have a bugle-shaped head, slender shanks, and thin, sharp threads that bite aggressively into wood. They have tremendous holding power, and in many cases they can be driven into wood or sheet metal without having to drill a pilot hole first. Wallboard screws come with Phillips or Robertson (square-drive) heads, and they are meant to be driven with power screwdrivers. The Phillips type is most common, but square-drive heads offer

a more positive grip and are less likely to strip out.

The trim-head screw has a very small head that can be countersunk below the surface. It is handy when you want an inconspicuous fastener. These are often used to attach trim to metal studs.

Galvanized or specially coated wallboard screws are great for wood decking, trim, and siding. One variation, the deck screw, has a specially shaped tip that cuts through the wood ahead of the threads to reduce splitting.

Bolts

Bolts are used for heavy-duty connections. Lag screws (sometimes called lag bolts), machine bolts, and carriage bolts are the most common

types found on construction sites. Use washers under bolt heads and nuts except when fastening metal hardware to wood.

Lag Screws

These heavy-duty fasteners look like big wood screws with hexagonal or square heads. They're used to fasten ledgers to walls, posts to beams, and where a connection can be made only from one side. Always drill pilot holes for lag screws. If the head needs to be sunk below the wood surface, drill a countersink hole first. Drill the pilot hole for the shank in two stages. First, drill as deep as the unthreaded portion of the shank, using a drill bit the same diameter as the shank. Then, drill the rest of the way

with a bit that's ⅛ inch smaller. Lubricate the threads with wax or soap before installing.

Machine Bolts

This type is used wherever an extrastrong connection is required. Drill the bolt hole with a bit that's 1⁄16 inch larger than the bolt diameter. If the pieces to be fastened are not too thick, you can tack them together and drill through both at once; for very long holes, you may have to drill one piece first, then use it to mark the second piece for drilling.

Carriage Bolts

Carriage bolts are similar to machine bolts, but have a dome-shaped head. This makes for a neat connection that is tamperproof from one side, so carriage bolts are frequently used for exposed hardware on doors and gates. Carriage bolts have a short, square shoulder under the head, which bites into the wood to prevent the bolt from turning while the nut is tightened. Holes for carriage bolts should be drilled the same diameter as the bolt shank. A washer isn't required under the head, but one should be used under the nut.

Stove Bolts

These bolts, with shanks that are completely threaded, are often used when fastening metal components together or to wood. Flat-head stove bolts can be set flush in wood or metal; round-head bolts accept a washer to prevent the head from biting into the surface.

Concrete Fasteners

Sometimes it is necessary to make connections to concrete or masonry. Attaching framing to foundations, furring strips to masonry walls, and cabinets to slab floors are typical examples. The fasteners listed below cover most applications.

Expansion Shields

The most common fastening method for concrete is an expansion shield and a lag screw. Drill a hole into the concrete with a rotary hammer. Place the shield in the hole. Insert the bolt through the material being fastened and into the shield. As you tighten the bolt, the shield will expand in the hole, making a tight connection.

Plastic Anchors

These miniature versions of the expansion shield are designed to be used with wood screws instead of lag screws. They are installed the same way as an expansion shield, and they're used for installing door thresholds to slabs, anchoring base cabinets, and other light-duty anchoring tasks.

Anchor Bolts

These can be used with concrete or masonry. Drill a hole the same diameter as the anchor and at least ½ inch longer than the depth the anchor will penetrate. Clean any dust from the hole. Place a nut and washer on the anchor and drive in the anchor. Tighten the nut to expand the base. The anchor is at its full holding power only when it tightens without rotating.

Powder-Actuated Fasteners

These hardened steel concrete nails are actually fired into the concrete with a shell that resembles a blank gun cartridge. They are fired by stud guns, which are expensive but can be rented. Low-velocity guns, which use a piston to drive the nail, are the safest type. If you have a lot of fastenings to make—installing furring strips on a basement wall, for instance—a stud gun can be a great time-saver. Choose fasteners that will penetrate the concrete 1 to 1½ inches.

Caution: Stud guns can be very dangerous if improperly used. Always read the safety instructions that come with the gun and follow them scrupulously. If you rent one, insist on a copy of these instructions. Never use a stud gun without ear and eye protection.

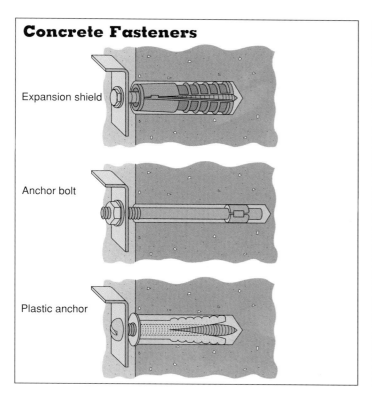

Concrete Fasteners

Expansion shield

Anchor bolt

Plastic anchor

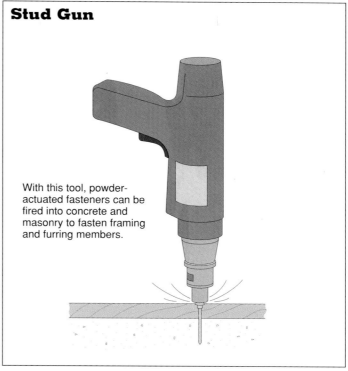

Stud Gun

With this tool, powder-actuated fasteners can be fired into concrete and masonry to fasten framing and furring members.

TOOLS

You don't need a lot of fancy equipment to do good carpentry. For proof of this, one need only look at the medieval timber-frame buildings of Europe, the eighteenth-century temples of Japan, or the colonial-era buildings of the United States—all of which were built by carpenters with the most rudimentary of tools. How you use the tools you have is more important than how many you own. Here's an introduction to the selection and use of the essential tools you'll need for basic carpentry tasks.

A tool belt or apron is essential for keeping your most-used hand tools where you can reach them. The tools visible in the near pocket of this belt are, from left to right, a combination square, framing hammer, nail set, cold chisel, wood chisel, pencil, torpedo level, and tape measure. Other tools convenient to have in your belt are a chalk line, bevel square, plumb bob, rasping plane, screwdriver, cat's-paw, pliers, and—needed sooner or later—tweezers.

BEGINNING WITH THE BASICS

Start your collection with a basic set of hand tools, an electric drill, and a portable circular saw. Try renting the tools you need but don't have—it's a good way to test them out. Buy the best tools you can afford—replacing an ineffective or broken tool often ends up costing more in the long run. Keep tools sharp and clean, and they will last a lifetime.

Measuring and Marking Tools

The tools used for measuring and marking are essential for every carpenter. If you can lay out your work accurately, you are well on the way to building projects that are sound, sturdy, and good-looking.

Tape Measures

These are available in lengths from 8 to 100 feet or more. For general-purpose work, a retractable tape measure with a wide (1-inch) blade is most useful. The wider blade extends 6 or 7 feet without sagging, giving you a longer reach when working alone. A 12-foot tape is fine for small jobs. A 25- or 30-footer is better for big projects and all-around use. Long tapes in lengths of 50 to 100 feet are used for laying out large foundations and locating buildings within property lines.

When you mark a dimension from a tape measure, use a V-shaped mark—carpenters call it a crow's foot), with the point of the V on the measurement you want. This is more accurate than simply drawing a line on a board, because the line is never quite straight and it's hard to tell exactly which

point on the line represents the correct dimension.

Wing Divider

A wing divider equipped with a pencil is useful for copying irregular lines. For example, if you need to fit a long board against an irregular rock chimney, you can hold the board beside the chimney and follow the ins and outs of the rocks with the point while the pencil copies that line on the board.

Combination Square

This essential tool is used to lay out 90- and 45-degree angles, and to check cuts for squareness. The blade can be removed and used as a ruler. Most models have a spirit level built into the head for small-scale leveling. A combination square can also be used to mark a line parallel to the edge of a board. First set the blade to the desired dimension, then hold a pencil against the end of the blade and slide the square along the edge to draw the line.

Framing Square

Of all the tools used by carpenters, the framing square is probably the most versatile. It can be used to mark 90-degree angles across wide boards, to lay out stud locations on wall

plates, and to check that assemblies are square. It's also used to lay out angled cuts for rafters and stairs (see page 73 for rafter layout; see page 80 for more on stairs). The tables stamped on the sides make it easy to calculate rafter lengths, lay out octagons, and figure the lengths of diagonal braces. There's also a table for calculating board feet, although most people find it easier to use a pocket calculator for this job.

Bevel Square

This tool has an adjustable, pivoting blade that can be locked in position with a wing nut. It can be set to duplicate any angle and transfer it to a board you want to cut. Use it whenever you need to make repeated angle layouts, as when marking cuts for a set of rafters.

This tool box is based on a classic design that carpenters have used for generations. It is long enough to house a 24-inch framing square, yet compact enough to be easily carried. For a construction diagram and dimensions for building this box, see page 40.

Measuring and Marking Tools

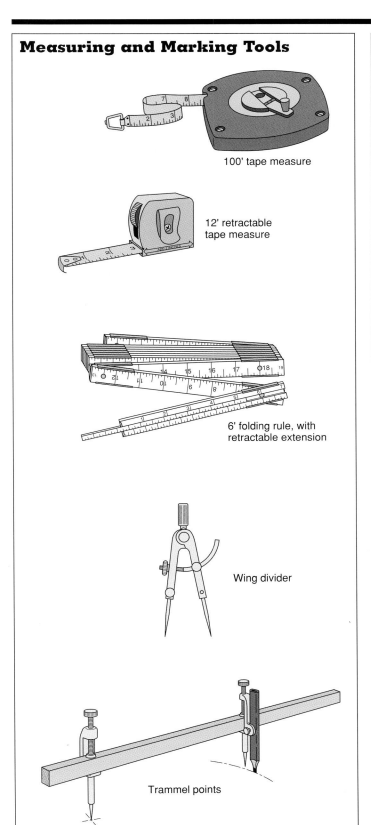

100' tape measure

12' retractable tape measure

6' folding rule, with retractable extension

Wing divider

Trammel points

Squaring Tools

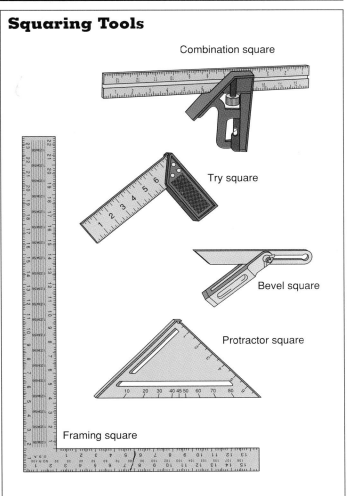

Combination square

Try square

Bevel square

Protractor square

Framing square

Chalk Line

This tool consists of a reel of string inside a container filled with colored chalk powder. It is used to lay out long, straight lines. In use, the chalk-covered string is stretched taut between two points, then plucked to produce the line. Use a chalk line to lay out wall locations on a floor, to mark long cuts on sheets of plywood, and to lay out a row of fasteners on floor, wall, and roof sheathing.

Levels

The several types of levels on the market all serve a common purpose: to give the builder a dependable reference for true level and vertical lines. Some types give accurate readings over just a few feet; others can reach out for distances of 100 feet or more.

Spirit Levels

These levels use a bubble in a vial of alcohol to establish plumb and level. A 24- or 30-inch carpenter's level is adequate for most work around the house, but if you are doing

Levels

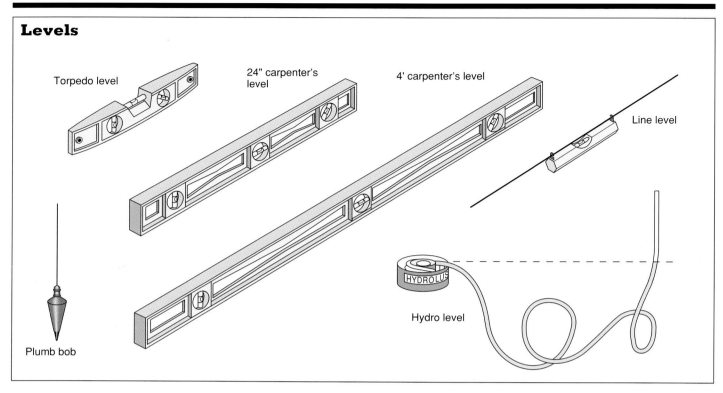

Torpedo level

24" carpenter's level

4' carpenter's level

Line level

Plumb bob

HYDROLU

Hydro level

extensive framing or working with brick or concrete block, you should also have a 4-foot level. You can extend the reach of a short level by taping it to a long, straight piece of wood. Most levels are made of aluminum extrusions, but masons prefer wood levels because they can be laid on a course of bricks or concrete blocks and tapped to make fine adjustments.

The torpedo level, which is about 12 inches long, fits in a tool belt. It's handy for plumbing work and setting concrete forms. The line level is an inexpensive, small spirit level that is suspended from a taut string. It can be used to level foundations or fences, but not with a high degree of accuracy. For best results, the level should be placed in the exact center of the string.

Hydro Level

Used primarily for leveling foundations, the hydro level is a simple tool that is always dead-on accurate. Hydro levels work on the principle that water always seeks its own level. A container of water is connected to a hose; on the other end of the hose is a clear tube that shows the level of the water inside. No matter where the hose is moved, the water in the tube remains at the same elevation. In order for a water level to function properly, care must be taken to eliminate air bubbles in the hose. A bit of food coloring in the water makes the level easier to read.

Plumb Bob

This instrument takes advantage of the immutable laws of gravity. Its history goes back at least to ancient Egypt.

Basically, it is nothing more than a pointed weight hung from a string. A plumb bob is the best tool for transferring a point on batter-board string lines to a spot on the ground, and it's also the easiest way to plumb a wall.

When holding a plumb bob by hand, maintain a firm stance and hold your elbow against your body to keep the bob steady. To plumb a wall, suspend the bob a couple of inches away from the wall surface, then measure back from the string at the top and bottom. If you are working outdoors, use a plumb bob during a time of day when the wind isn't blowing—it's difficult to get a good reading in even a light breeze.

Builder's Level

Sometimes incorrectly called a transit, this is the tool professionals use to establish a true level plane for foundations. It is a swiveling telescope mounted on a tripod. A set of crosshairs in the scope allows the user to read elevations accurately at distances of 50 feet or more. Although the cost of this instrument makes it impractical for a homeowner to buy, it can be rented.

It takes two people to use a builder's level. One person holds a tape measure or surveyor's rod against some reference point, while the other person notes the dimension visible in the crosshairs of the level. The tape is then moved to the next spot (usually a building corner) and raised or lowered until the same dimension shows in the crosshairs.

TOOLS FOR CUTTING AND SHAPING

The tools used to cut, bore, and shape wood take many forms, but you can do most jobs with just a few of them. Remember to keep cutting tools sharp. Not only does a dull tool make your job harder, but it can slip and injure you as well.

Handsaws

Although most of the work carpenters once did with handsaws is now done with power tools, the handsaw still has a place in every carpenter's toolbox. It can reach into the corners of a notch, make cuts flush with an adjoining surface, and get into other places where a power saw can't go.

Panel Saws

There are two basic types of panel saws. Crosscut saws are designed to cut across the grain of the wood. They have teeth with beveled tips that score the wood fibers on both sides of the cut. Crosscut saws have between 8 and 12 teeth per inch; an 8- or 10-point saw is a good choice for general-purpose work. Ripsaws have teeth sharpened straight across; they have 5 or 6 teeth per inch and are designed to cut with the grain. If you own a power circular saw, you are not likely to need a ripsaw.

It takes some practice to get the hang of using a handsaw. Most beginners run into trouble because they hold the saw too tightly and bear down too hard. This invariably makes the saw wander off the cut line, causing binding and hopelessly out-of-square cuts.

A good exercise to try when learning to use a handsaw is to draw a series of perpendicular lines on a piece of 2×4, then saw off slices one after another. Check each cut for squareness before making the next. You may find it helpful at first to square a line completely around the board; with practice you'll be able to make a good cut with just a line across the wide face.

Compass Saws

These saws are handy for starting cuts after first drilling a hole, making plunge cuts in fiberboard or wallboard, and hand cutting a gentle radius. The keyhole saw is a smaller version of the compass saw.

Hacksaws

Hacksaws are designed primarily for cutting metal, although they also work well for plastics. Carpenters use them for cutting metal door thresholds, weather stripping, and

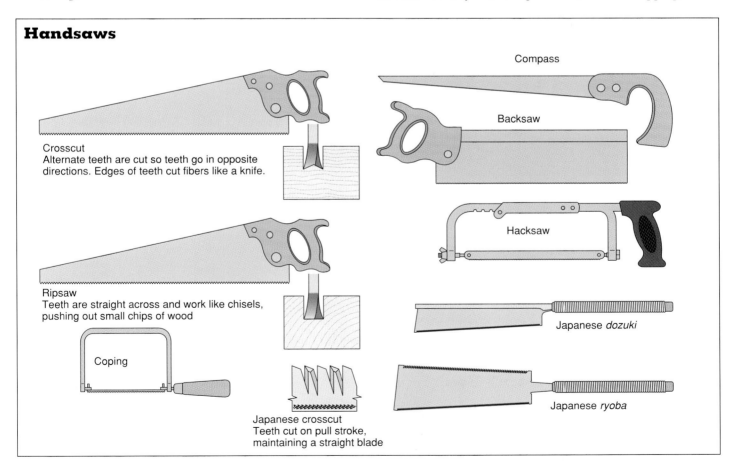

Handsaws

Compass

Backsaw

Crosscut
Alternate teeth are cut so teeth go in opposite directions. Edges of teeth cut fibers like a knife.

Hacksaw

Ripsaw
Teeth are straight across and work like chisels, pushing out small chips of wood

Coping

Japanese *dozuki*

Japanese crosscut
Teeth cut on pull stroke, maintaining a straight blade

Japanese *ryoba*

Handsaw Techniques

1. Grasp saw *loosely*. Hold piece of scrap against cut line as a guide, and draw saw backward once or twice to start cut. Don't use thumb to start cut—if the saw skips, you can cut through the tendon in your knuckle.

2. Begin sawing with light, smooth strokes. Don't push down on saw—a sharp blade cuts with its own weight. Be sure cut follows line exactly, from beginning to end; if you try to correct for a wayward cut, the blade will bind up.

3. When you near end of cut, reach across with your free hand to support cutoff piece so it won't fall and splinter good piece.

reinforcing bars for concrete. Blades are available with 14 to 32 teeth per inch. Coarse teeth make for a faster cut, but catch on thin materials. Select a blade that will have at least two teeth in the material at all times.

Backsaws

These saws have fine teeth and a reinforced steel spine that stiffens the blade. They are used where smooth, accurate cuts are necessary, as in furniture and cabinet joinery. A smaller version, the dovetail saw, can be handy for trimming small moldings.

Coping Saws

Coping saws have a narrow blade that is held in tension by a steel frame. Their primary function is to make coped cuts for moldings that meet at an inside corner (see page 88 for more on coped moldings). They can also cut circles and odd-shaped holes in the middle of a board, within the limits of the depth of the frame.

Japanese Handsaws

Japanese-style saws have recently become very popular among finish carpenters. They are unlike Western saws in that the teeth point backward (toward the user), so the saw cuts on the pull stroke. The blades are very thin and have elongated, razor-sharp teeth. These saws cut quickly and leave a smooth surface with very little splintering. Because the delicate teeth are difficult to sharpen, most Japanese-style saws sold in the United States have replaceable blades.

Two types of Japanese saws are of interest to Western carpenters. The *ryoba* saw has teeth on both sides of the blade. One set of teeth is for crosscutting; the other set is for rip cuts. The *dozuki* saw looks much like a dovetail saw, and it's used for the same kind of precision work.

Hybrid Saws

One interesting variation of the handsaw is a hybrid that combines a Western-style body with teeth of the Japanese style. These saws have a short, stiff blade that cuts on the push stroke. The teeth cut very aggressively—these saws are among the fastest cutting available. The cut is rather coarse, so hybrid saws are best suited to rough work.

Miter Boxes

A miter box is a device that holds a saw steady for making clean and accurate cuts. Several variations of miter boxes are available. All are used to make accurate cuts for moldings. The simplest type is made of hardwood or plastic; it is limited to 45- and 90-degree miters in narrow stock. Fancier miter boxes can be adjusted to any angle between 45 and 90 degrees. Some of these use a backsaw to make the cut; others use a bowsaw, a saw that looks like a hacksaw, with a narrow blade in a frame. The latter type provides the best control and truest cut. It is also considerably lighter than a backsaw, so many people find it easier to use.

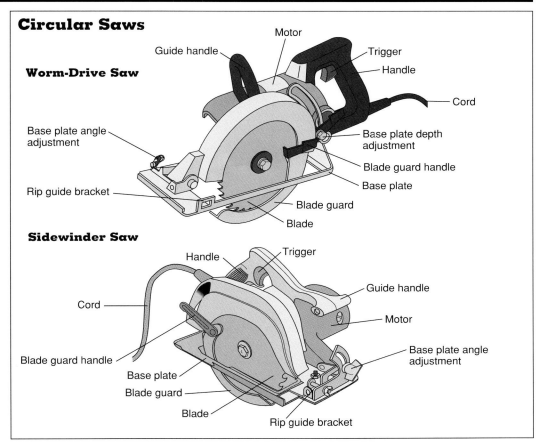

Circular Saws

The circular saw is the carpenter's workhorse. It can crosscut, rip, miter, and bevel much faster than a handsaw. It can make cuts in the middle of a board. With the right blade, a circular saw can cut almost any material, including concrete and steel.

Choosing a Circular Saw

Circular saws are designated by the size of the largest blade they will accept. These sizes range from tiny 4-inch saws all the way up to big 16-inch beam saws. The standard carpenter's saw uses a 7¼-inch blade, which is large enough to cut 2-by lumber at a 45-degree angle or to cut 4-by lumber in two passes.

Circular saws come in two basic configurations. Sidewinder saws have the motor mounted perpendicular to the saw blade, on the left-hand side of the saw. Most homeowner saws are of this type. Some right-handed people find these saws awkward to use because they have to lean across the saw in order to watch the blade while making a cut. Worm-drive saws have the motor mounted parallel to the blade, on the right-hand side of the saw, giving right handers a better line of sight. These saws are usually heavy-duty models and are more expensive than the sidewinder variety.

Horsepower ratings given by manufacturers for their saws can be misleading, because some list the maximum horsepower the motor can produce, even if the saw would burn up if used at this level for more than a minute or two. A better gauge of power is the ampere rating, which gives the motor's maximum current consumption under continuous use. This information is listed on a label attached to the saw body. A circular saw rated at 10 to 13 amps should be powerful enough for any job you are likely to encounter around the home.

Circular Saw Techniques

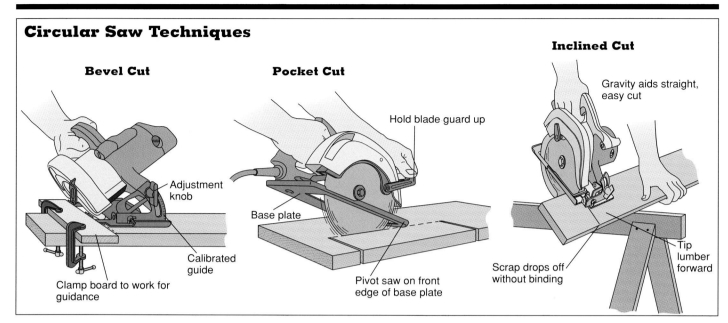

Bevel Cut

Adjustment knob

Calibrated guide

Clamp board to work for guidance

Pocket Cut

Hold blade guard up

Base plate

Pivot saw on front edge of base plate

Inclined Cut

Gravity aids straight, easy cut

Scrap drops off without binding

Tip lumber forward

Using the Circular Saw

Although it is a powerful and versatile tool, the circular saw is no better than the person using it. It takes some practice to get the most out of your saw. Remember that the circular saw cuts upward, which will cause some splintering on the top face of the piece being cut. For best results, keep the best face of the wood down, especially when cutting plywood.

One common mistake made by beginners is to feed the saw too slowly into the cut. When this happens, the teeth stop cutting and instead start to rub against the end of the cut. This generates enough heat and friction to dull the saw teeth in seconds. Always try to maintain a smooth, even rate of feed at all times.

Crosscuts

Trimming boards to length is the staple diet of the circular saw. To start the cut, place the tip of the base plate on the work and align the blade with the cut line. Do not let the blade touch the wood until the motor is running. Feed the saw into the cut, remembering to stay to the waste side of the line. Once the cut is complete, release the trigger and wait for the blade to stop spinning before you lift the saw.

Rip Cuts

When you buy a circular saw, get a rip fence to go with it. This accessory makes it much easier to cut a straight line parallel to the edge of a board. Adjust the fence to the required setting and make the cut as you would a crosscut. If the saw blade tends to bind, jam a screwdriver or wood wedge in the kerf (cut created by saw blade) to keep it open. If the board is too wide to use the rip fence, clamp a straight-edge next to the saw to guide it. This technique is especially useful when cutting plywood.

Basic Saw Cuts

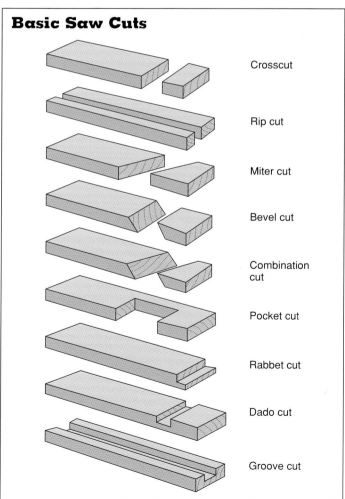

Crosscut

Rip cut

Miter cut

Bevel cut

Combination cut

Pocket cut

Rabbet cut

Dado cut

Groove cut

Long-Cut Techniques

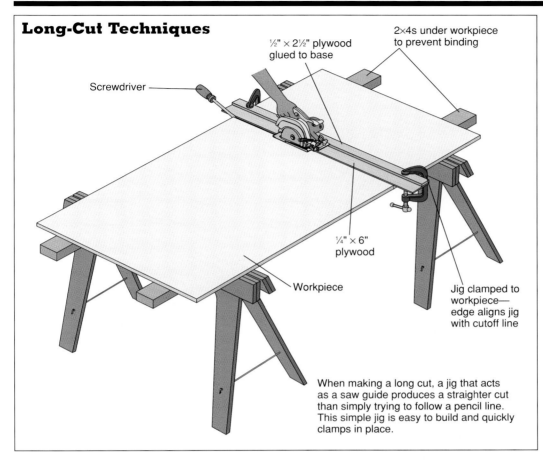

Screwdriver

½" × 2½" plywood glued to base

2×4s under workpiece to prevent binding

¼" × 6" plywood

Workpiece

Jig clamped to workpiece— edge aligns jig with cutoff line

When making a long cut, a jig that acts as a saw guide produces a straighter cut than simply trying to follow a pencil line. This simple jig is easy to build and quickly clamps in place.

Bevel and Miter Cuts

A circular saw has a base plate adjustment angle attached to the base plate at the front of the saw, calibrated from 0 to 45 degrees. To make angled cuts, loosen the knob, tilt the base plate to the desired angle, and tighten the knob. (On less expensive saws, the quadrant is often inaccurate; check the angle with a protractor square.) To make accurate bevels, start the cut on the line, then clamp a board next to the saw to guide it straight across.

Pocket Cuts

Sometimes it is necessary to start a cut in the middle of a board. To make a pocket cut, rest the front end of the base plate of the saw on the work,

hold the blade guard up by its lever, and align the blade with the cut line. With the blade just above the material to be cut, start the saw and slowly lower it into the wood. When the base plate rests firmly on the work, move the saw forward to complete the cut. Never back the saw up in a pocket cut; the teeth at the back of the blade will catch, causing the saw to leap out of the cut, ruining your work and possibly injuring you in the process.

Dado and Rabbet Cuts

These two cuts can be made quickly with a circular saw. Set the saw blade to the required depth and cut the outer edge first. Then make several

passes through the center portion to remove most of the waste. Trim away what is left with a chisel.

Circular Saw Safety

The circular saw is responsible for more job-site injuries than any other tool. Always follow the following safety precautions when using this tool.

• Never remove or wedge up the blade guard. It is designed not only to save fingers but to allow you to put the saw down while the blade is still spinning.

• Always support the work securely and in such a way that it won't pinch the blade as the cut is completed.

• Unplug the saw before changing blades or making depth and angle adjustments.

• Wear safety glasses, especially when you are using abrasive blades.

Saw Blades

Using the right blade is at least as important as the quality and horsepower of your saw. Circular saw blades should always be kept sharp. A dull blade won't cut cleanly, and it is more likely to kick back and cause injuries. Choose the right blade for the job, and your work will be the better for it.

Carbide or Steel?

Saw blades for circular saws come with steel or carbide-tipped teeth. Most new saws come with steel blades. These will work fine, but you can expect to get only six to eight hours of use from them before they need resharpening—even less if you are cutting materials that contain abrasive glues, such as plywood or particleboard. Carbide-tipped blades cost more, but they last up to 50 times longer than steel. With normal use, they may never need sharpening.

The number of teeth on a standard 7¼-inch circular saw blade varies from 12 to 60 or more. In general, more teeth give a smoother but slower cut. For framing work and wet lumber, use a blade with 12 to 24 teeth. Use finer blades for finish work.

Carbide saw blades come in different thicknesses. Thin-kerf blades are about ⅟₁₆ inch thick, usually with alternate-top-bevel (ATB) teeth. They cut almost without effort. Standard-thickness blades,

Folding Sawhorse Project

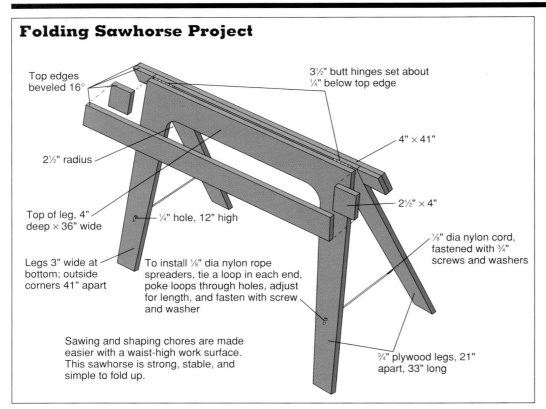

Top edges beveled 16°

3½" butt hinges set about ¼" below top edge

4" × 41"

2½" radius

2½" × 4"

Top of leg, 4" deep × 36" wide

¼" hole, 12" high

⅛" dia nylon cord, fastened with ¾" screws and washers

Legs 3" wide at bottom; outside corners 41" apart

To install ⅛" dia nylon rope spreaders, tie a loop in each end, poke loops through holes, adjust for length, and fasten with screw and washer

¾" plywood legs, 21" apart, 33" long

Sawing and shaping chores are made easier with a waist-high work surface. This sawhorse is strong, stable, and simple to fold up.

---

about ⅛ inch thick, are available in all tooth configurations. Stiffer, they produce smoother cuts, but require more effort from both saw and operator.

Saw Blade Types

Combination Blade
This type, also known as a general-purpose blade, should be the first saw blade you buy. Most models have a tooth pattern of two or four ATB teeth, followed by a raker tooth with a flat-top grind. A combination blade does a decent job making crosscuts, rip cuts, miters, and bevels, although it won't cut quite as well as specialized blades.

Crosscut Blades
These blades have teeth with an ATB or triple-chip grind. They are designed to make smooth crosscuts with a minimum of splintering. ATB blades work best for solid woods; triple-chip blades are excellent for cutting plywood, particleboard, and laminates.

Rip Blades
Rip blades have a small number of teeth with a flat-top grind and an aggressive hook angle. They make cuts along the grain faster than any other blade type, but cause splintering if used for crosscutting.

Abrasive Blades
These are used for cutting concrete, masonry, and steel. They're composed of grains of abrasives bound together with fibers. Masonry- and steel-cutting blades are not interchangeable, so make sure that the blade you buy is designed for the material you'll be cutting.

Make cuts in masonry and concrete ¼ inch at a time. These materials make lots of dust, so always wear a dust mask. You can protect the saw motor from abrasive particles by taping a dust mask over the air intake vents. Make cuts in steel with the blade set at full depth. Sparks will fly, so be sure to wear safety glasses.

Planes and Chisels

Planes and chisels are simple tools, but few people know how to use them well. Planes must be tuned up and properly adjusted, and both planes and chisels have to be kept sharp. Don't scrimp on quality when buying these tools—cheap ones will never work the way they should.

Planes

Block Plane

The block plane is the first plane you should buy. It's small enough to be held in one hand, and it is indispensable for fitting joints and rounding off sharp edges. Unlike the larger bench planes, the blade of the block plane fits into the body with its bevel up. This allows the plane to shave end grain without chattering—a quick skipping of the blade that will serrate the wood.

Smoothing Plane

This plane has a relatively short sole (about 9 inches) that allows it to follow the contours of a board when planing. A properly tuned smoothing plane can take a shaving as fine as one thousandth of an inch, leaving a glassy surface that is smoother than any sandpaper could make it.

Fore Plane

The fore plane, or jack plane, is used to dimension lumber. It is 14 to 15 inches long and, like other bench planes, has the blade set at about a 45-degree angle.

Jointer Plane

The jointer plane is used for straightening the long edges of boards and for planing doors. It is ideal for achieving a tight joint on edge-joined surfaces.

Using Hand Planes

Even new planes require some tuning up before they are ready to use. Check the sole for flatness. Place a straightedge against the bottom of the plane, hold the

Using a Plane

Planing With the Grain

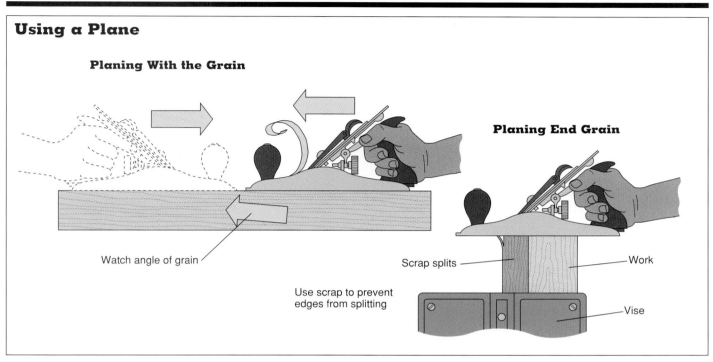

Watch angle of grain

Use scrap to prevent edges from splitting

Planing End Grain

Scrap splits

Work

Vise

Smoothing Plane

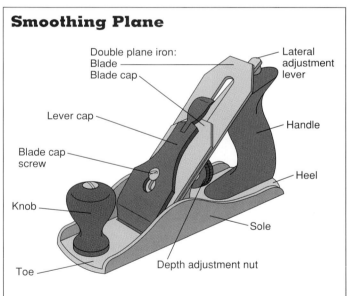

Double plane iron:
Blade
Blade cap

Lever cap

Blade cap screw

Knob

Toe

Lateral adjustment lever

Handle

Heel

Sole

Depth adjustment nut

Planes

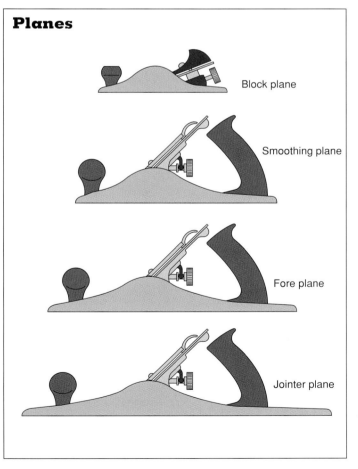

Block plane

Smoothing plane

Fore plane

Jointer plane

plane up to the light, and look for any sign of light shining between the straightedge and the plane. Remove high spots from the bottom of the plane with a mill file, rechecking frequently as you work. Check the blade cap to see if it fits tightly to the blade; if it does not, shavings will catch under its edge and clog the plane. Use a file to true up the edge of the blade cap until it fits snugly. Finally, hone the blade (even a new one) until it is sharp enough to shave with.

To use the plane, first adjust the blade cap. For

Sharpening Planes and Chisels

Planes and chisels require frequent sharpening. Buy a sharpening stone and learn how to use it. With practice, you can hone an edge and be back at work in a couple of minutes.

Sharpening stones come in various degrees of coarseness, depending on the level of finish required. Buy a combination stone, which has a coarse-grit abrasive on one side (for preliminary sharpening) and a fine-grit abrasive on the other (for final honing). Always lubricate the stone with oil or water, depending on the type of stone you buy, to prevent clogging the surface with particles of metal.

The correct sharpening angle for plane and chisel blades is about 30 degrees, which makes the width of the bevel twice the thickness of the blade. Sharpen the bevel only; the back of the blade should be kept absolutely flat.

Hold the blade on the stone and rock it back and forth until you feel the bevel seat against the stone; keep the blade at that angle and push it back and forth until you can feel a slight burr raised up on the flat side. This is called a wire edge. Then turn the blade over and hone the wire edge off.

A very dull or nicked edge may require regrinding. Use a belt sander or a bench grinder, and dip the tool frequently in water to keep the blade from overheating and losing its temper. Then hone the blade on the sharpening stone.

rough work, set it about ¹⁄₁₆ inch behind the cutting edge; for fine planing, it should be only ¹⁄₆₄ inch back. Secure the blade in the plane body with the lever cap. Adjust the depth of the cut by turning the depth adjustment nut, sighting down the sole until you see the blade protrude slightly. Test the plane on a piece of scrap wood and make fine adjustments. If the blade seems to be cutting more on one side, move the lateral adjustment lever until the shavings come from the center of the blade.

Always plane with the grain of the wood (on some pieces you may have to change direction in the middle of the board to avoid tear-out). Start the stroke by applying downward pressure to the front of the plane; as you get to the end of the piece, shift pressure to the rear to avoid rounding off the ends.

Chisels

Every carpenter needs one or two chisels in the toolbox. Chisels are used for chopping waste wood from notches, trimming joints, and mortising for door hardware. Like planes, they must be kept sharp to work effectively. Never use a chisel as a pry bar or a paint-can opener.

Choosing Chisels

Chisels come in many shapes and sizes. Butt chisels, which have a short blade, are the

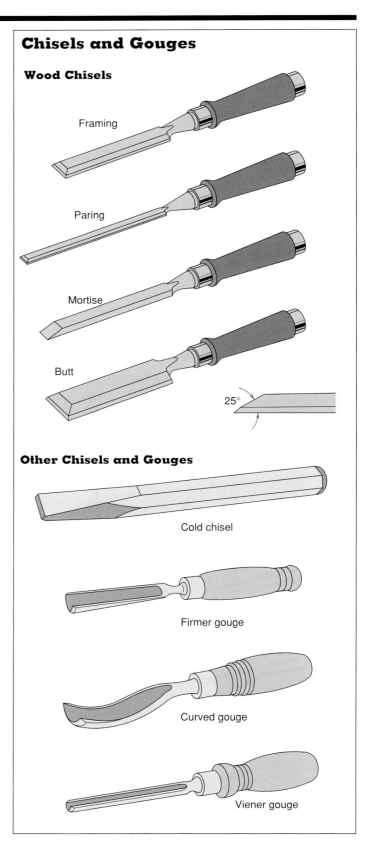

Chisels and Gouges

Wood Chisels

Framing

Paring

Mortise

Butt

25°

Other Chisels and Gouges

Cold chisel

Firmer gouge

Curved gouge

Viener gouge

easiest to control. They range in width from ¼ to 2 inches. A 1-inch chisel is a good choice for general-purpose work. Wider chisels (1½ or 2 inches) are more useful for rough framing; the smaller sizes are used mostly for cabinet joinery.

Other types of chisels are available for specialized applications. Paring chisels are designed to be pushed by hand to take fine shavings when trimming joints. Mortise chisels have narrow, stout blades for chopping the deep holes required for mortise-and-tenon joints. Framing chisels are designed for rough work such as chopping away a portion of 2×4 framing to fit around rough plumbing. Cabinet-maker's chisels have longer blades that can be used for both paring and light chopping. Cold chisels, which should be part of every carpenter's tool kit, are invaluable for chopping off rusted bolts and other heavy-duty tasks.

Using Chisels

For rough work when you have to remove a lot of material, use the chisel with the bevel against the work. Don't try to cut too deeply all at once. Chop down about ⅛ inch, remove the waste, and then cut deeper as required.

For finer cutting, as when shaving a joint to fit, keep the flat side of the chisel down. Most light cuts can be made with hand power alone. Place the end of the chisel handle in the palm of your hand, and lock your elbow tightly against your body. Make the cut by shifting your weight, using your whole body to power the cut. You'll find that

this method gives you much better control over the blade than pushing the chisel with your arms.

Drills and Bits

Carpentry work often requires drilling holes, particularly for installing screws and bolts. There are several types of drills and many more kinds of bits for different drilling tasks.

Hand Drills

There isn't much call for hand drills these days, but there are two types you may find useful. The "eggbeater" drill is powered by a crank on the side of the tool. It can drill holes up to about ¼ inch diameter in metal and ½ inch in wood. The bit brace has an offset handle and a special chuck designed to hold an auger bit with a square tang. You can apply quite a bit of torque with a brace, enough to drill holes up to about 1½ inches in diameter.

Power Drills

If you can afford only one power tool, buy a power drill. It can drill a hole in a fraction of the time it would take with a hand drill, and most power drills can be used to drive screws as well.

Power drills are classified according to the largest size drill-bit shank the chuck will accept. These sizes are commonly ¼ inch, ⅜ inch, and ½ inch. Smaller drills turn at higher speeds than larger sizes. High rpm works best for small drill-bit sizes; a drill that spins at a slower speed has more muscle for turning big

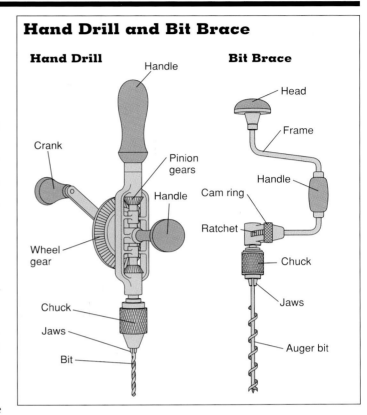

Hand Drill and Bit Brace

Hand Drill
Handle
Crank
Pinion gears
Handle
Wheel gear
Chuck
Jaws
Bit

Bit Brace
Head
Frame
Handle
Cam ring
Ratchet
Chuck
Jaws
Auger bit

bits. A ⅜-inch drill is a good choice for general-purpose work, but if you expect to do much drilling with auger bits or hole saws, you should consider a ½-inch model.

A drill with a variable-speed trigger and a reversing switch is well worth the extra cost. The variable-speed feature allows you to start at low rpm, then speed up to drill the hole. It is also useful for driving screws, when slower-than-normal speeds are needed. A reversing switch allows you to remove screws or back a drill bit out of a hole.

Using a power drill is quite simple. It usually helps to make some kind of starter hole, with an awl or the tip of a nail, to keep the bit from skipping around when you start the drill up. Apply moderate pressure to feed the bit into the

material. On deep holes, back the bit out periodically to clear out waste and cool the bit.

Cordless Driver-Drills

A cordless driver-drill with a ⅜-inch chuck drills holes up to ¼ inch in steel and ½ inch in wood. Better models have variable-speed and reverse features, as well as an adjustable clutch that stops the chuck from turning at a predetermined torque level.

Cordless driver-drills excel at driving screws. Most models can drive dozens, even hundreds, of screws on one battery charge. The adjustable clutch can be set so that screws won't strip out in the wood or be driven too deeply below the surface.

Drills

Power Drill

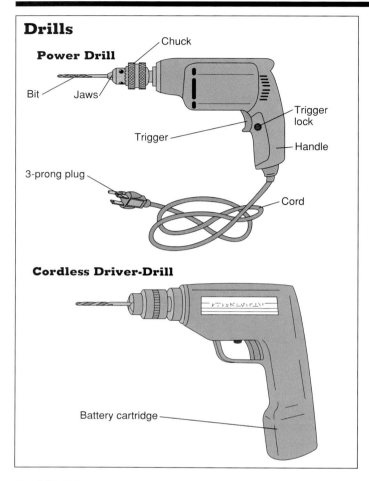

Chuck
Bit
Jaws
Trigger lock
Trigger
Handle
3-prong plug
Cord

Cordless Driver-Drill

Battery cartridge

Drill Bits

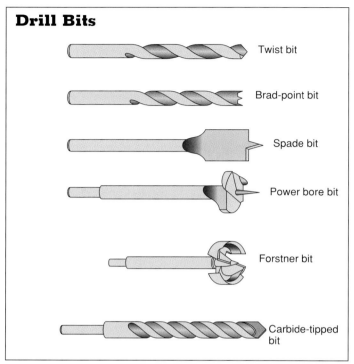

Twist bit

Brad-point bit

Spade bit

Power bore bit

Forstner bit

Carbide-tipped bit

Drill Bits

There are many types of drill bits available for different drilling jobs. Some of the most common types used by carpenters are listed below. Start your collection with a set of twist drills. Other bits are better purchased as the need arises.

Twist Drills

These bits can be used for drilling wood or metal. The best kind are made of high-speed steel, which stays sharp longer. High-speed bits are stamped HS or HSS on the shank. A set of twist drills in diameters from $\frac{1}{16}$ inch to $\frac{1}{2}$ inch, in $\frac{1}{32}$-inch increments, allows you to drill pilot holes for any screw or nail and for bolts up to $\frac{1}{2}$ inch in diameter.

If you use a twist drill in metal, be sure to apply enough pressure to keep the bit cutting. A drop of oil or cutting fluid on the tip of the bit makes the drilling go faster and reduces the chance of the bit overheating from friction.

Drills can be sharpened with a bench grinder, though for drills less than $\frac{1}{4}$ inch in diameter, it is just as easy to buy new ones. When sharpening, aim to match the factory grind.

Auger Bits

These bits are the best choice for drilling holes larger than $\frac{1}{2}$ inch in diameter. They work only in wood. Auger bits have a cone-shaped screw on the end that pulls the bit into the wood as it turns. Cutting spurs score the outside edge of the hole, while a cutting lip removes the waste from the middle.

Spade and Power Bore Bits

Spade bits are inexpensive, and they work quite well for drilling wood. They are available for drilling holes from $\frac{1}{4}$ to 2 inches in diameter. If you use one of these bits to drill a deep hole, clear out the chips frequently or the bit will get stuck. A power bore bit works in a manner similar to the spade bit, but gives a slightly finer cut. Sizes range from $\frac{1}{4}$ to 1 inch.

Brad-Point and Forstner Bits

Brad-point drills look like twist drills but have cutting spurs like those of an auger bit. They drill a very clean hole with a minimum of tearout on the wood surface. Brad-point bits are ideal for drilling holes for dowels or the wood plugs used to cover screw heads. Sizes range from $\frac{1}{8}$ to $\frac{9}{16}$ inch.

Forstner bits are the cleanest-cutting bits available. Because the bit is guided by its sharpened rim, it drills angled or overlapping holes without wandering. Forstner bits come in sizes from $\frac{1}{4}$ inch up to 3 inches. The larger sizes have saw teeth around the rim for faster cutting.

Carbide-Tipped Drill

The carbide-tipped twist drill is tipped with extrahard carbide steel and is used to drill concrete or stucco. Sizes range from $\frac{1}{4}$ to 1 inch.

ASSEMBLY TOOLS

Once you have cut, drilled, and made the parts for a project, you'll have to put them all together. The tools in this section are the ones you'll need to install—and remove, if need be—the various fasteners used in carpentry work.

Hammers

If you think a hammer is a simple tool, just check out the hammer display rack at any well-equipped hardware store. The variations seem endless. There are hammers with curved claws, straight claws, with no claws at all. There are hammers with wood handles, steel handles, fiberglass handles. There are hammers with heads made out of steel, plastic, wood, rubber, and rawhide. Some look like hatchets. And every one seems to come in 10 different sizes.

Don't despair—you can do 90 percent of the hammering you'll ever have to do with just one, or maybe two, hammers.

Claw Hammers

Start out with a 16-ounce claw hammer. The claws can be either curved or straight. Wood and fiberglass handles offer the best shock absorption and are the most comfortable to use. Hammers with solid steel handles are practically unbreakable, making them good for demolition work and other jobs involving a lot of prying. A 16-ounce hammer, which should have sharp claws and a smooth, slightly domed striking face, can be used for light remodeling, cabinetmaking, and finish carpentry.

If you will be doing any framing projects, make your second hammer a 20- to 28-ounce framing hammer, with ripping claws and a checkered face. The ripping claws are better than the curved type for prying and removing large nails; the checkered face will bite into the head of a nail to keep the hammer from glancing off. Framing hammers are powerful tools, capable of delivering a tremendous blow.

Special-Purpose Hammers

Depending on the kind of work you'll be doing, you may eventually want to add other types of hammers to your collection.

• Sledge hammers are good for demolition work and for nudging walls and beams into position on framing jobs. A short version, called a single jack, is handy for driving stakes for concrete forms.

• Wallboard hammers have an angled head for nailing into corners. The striking face is rounded so the head of the nail can be dimpled into the wallboard surface. One end of the head has a small hatchet blade for chopping out notches and cleaning up the cut edges of wallboard sheets.

• Shingling hatchets have an adjustable gauge built into the head for setting the exposure of shingles without

Hammers

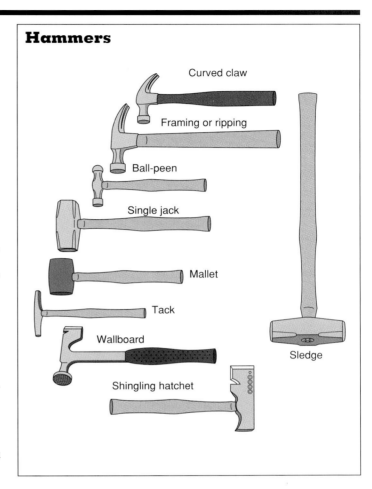

measuring. The end opposite the striking face has a blade (sometimes replaceable) for cutting asphalt shingles and roofing felt.

• Ball-peen hammers are designed for metalworking. The specially hardened head won't chip or shatter when it hits a hard surface.

• Mallets have heads made of wood, rubber, plastic, or leather. All are designed to deliver a solid blow without marring the surface. They are used to assemble cabinet joints and for striking wood-handled chisels and gouges.

• Tack hammers are lightweight and are magnetized at one end to set tacks otherwise too small to hold.

Screwdrivers, Wrenches, and Pliers

Every homeowner needs a selection of these simple tools around the house for installing and removing screws and bolts.

Screwdrivers

The two most common types of screwdrivers are the single-slotted (standard) and cross-slotted (Phillips) screwdrivers. In order to do even the most basic work, you need a medium-sized slotted and a No. 2 Phillips head. Add other sizes of shanks and heads to your tool collection as the need arises. Among the types

Screwdrivers

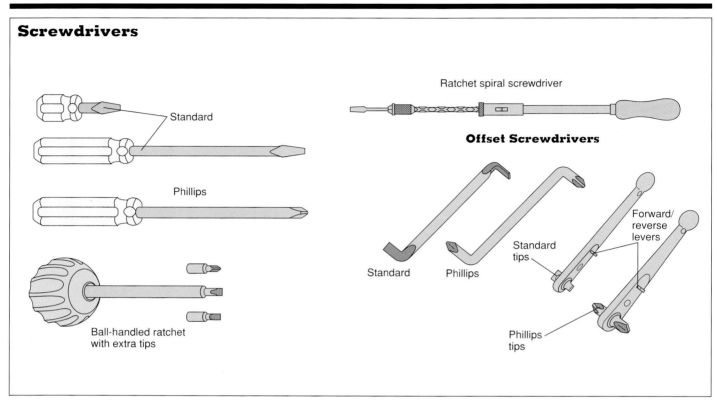

Standard

Phillips

Ball-handled ratchet with extra tips

Ratchet spiral screwdriver

Offset Screwdrivers

Standard

Phillips

Standard tips

Forward/ reverse levers

Phillips tips

Wrenches

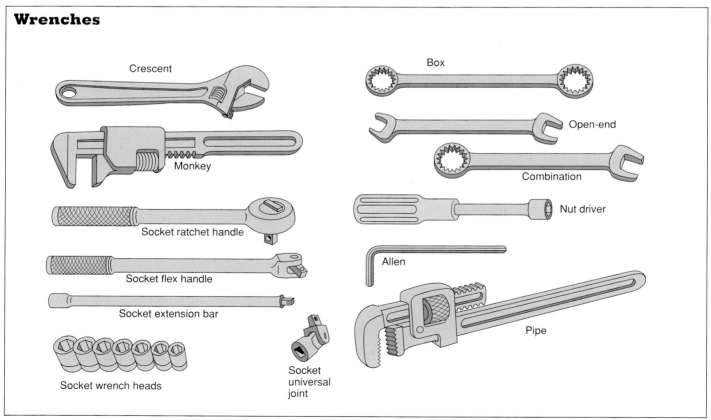

Crescent

Monkey

Socket ratchet handle

Socket flex handle

Socket extension bar

Socket wrench heads

Socket universal joint

Box

Open-end

Combination

Nut driver

Allen

Pipe

Pliers

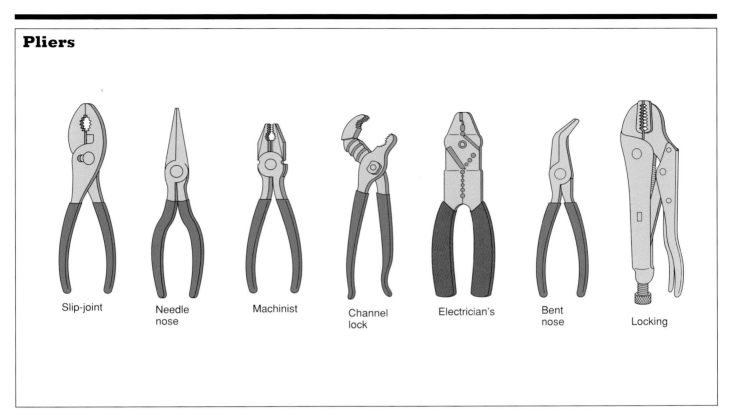

Slip-joint   Needle nose   Machinist   Channel lock   Electrician's   Bent nose   Locking

of manual drivers you can choose from are offset screwdrivers, ball-handled ratchets, ratchet spiral screwdrivers, and lever-style ratchets. All accept interchangeable tips.

If you have an electric drill or a cordless driver-drill, buy a magnetic bit holder that you can use for power-driving screws. Not only is it faster than driving screws by hand, but replacement tips usually cost less than a dollar, so it's easy to keep sharp, fresh bits on hand.

Wrenches

If you already have wrenches for automotive work, you won't need another set for household carpentry unless you want to keep them separate. Wrenches are used most often for bolting beams and mudsills, but they are also handy for plumbing and installing appliances.

An 8- or 10-inch adjustable crescent wrench is good enough for most carpentry jobs. Box, open-end, combination, and socket wrenches are better for getting into tight spots and are less likely to slip on stubborn nuts and bolts. Nut drivers are handy for the miscellaneous sheet-metal and hardware tasks that come with any remodeling project. Allen wrenches are essential for installing front-door locksets.

Pliers

Sooner of later you will need all the pliers shown, but start with a pair of machinist pliers and a pair of adjustable channel lock pliers. As ubiquitous as slip-joint pliers are, they are not nearly as useful as the other specialized types.

Needle nose, electrician's, and bent nose pliers assist with electrical work. A locking pliers is a rough-and-ready tool that can serve as a stand-in pipe wrench, clamp, or wrench.

Pry Bars and Nail Pullers

If you do any remodeling work around the home, you'll need one or more prying tools for disassembling old work. You'll want them for new work, too—as much as we hate to admit it, mistakes happen to the best of us.

Pry Bar

This tool is an essential item in every carpenter's toolbox. The best ones are made of spring steel and have slim, tapered ends for reaching into narrow

crevices. Use a pry bar to separate framing members, remove trim, and pull nails. You can prevent damage to surrounding areas if you slip a piece of scrap wood between the pry bar and the surface you're prying against.

Cat's-paw

The cat's-paw is used to pull nails that have been driven all the way down. It has stout claws that can be driven under the head of a nail to hook the head. When you buy a cat's-paw, look for one with sharp, pointed claws; a cat's-paw with rounded ends won't dig in deep enough to get a good hold on the shank of a nail.

End Cutters

Sometimes the head of a nail breaks off when you are trying to pull it. When this happens, a pair of end cutters lets you grab on to the shank of the nail and work it out. End cutters are also good for pulling finishing nails, whose heads are difficult to grasp with other tools. Helpful hint: If you're pulling finishing nails from a piece of trim you want to reuse later, pull the nail through the *back* of the piece; that way, you'll minimize damage to the face.

Wrecking Bar

This is a heavy-duty tool for serious demolition. It works well for removing old wood flooring, dismantling walls, and stripping concrete forms. It also pulls nails easily once the heads have been lifted clear of the surface with a cat's-paw.

Pry Bars and Nail Pullers

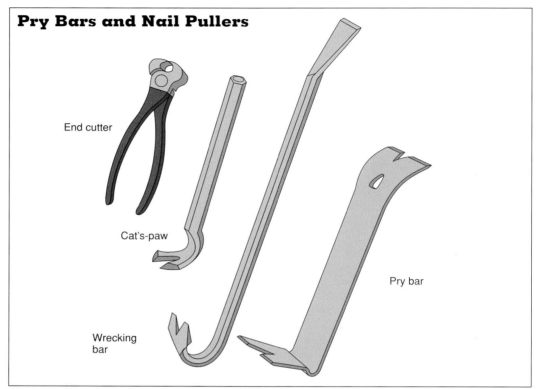

End cutter

Cat's-paw

Wrecking bar

Pry bar

Toolbox Project

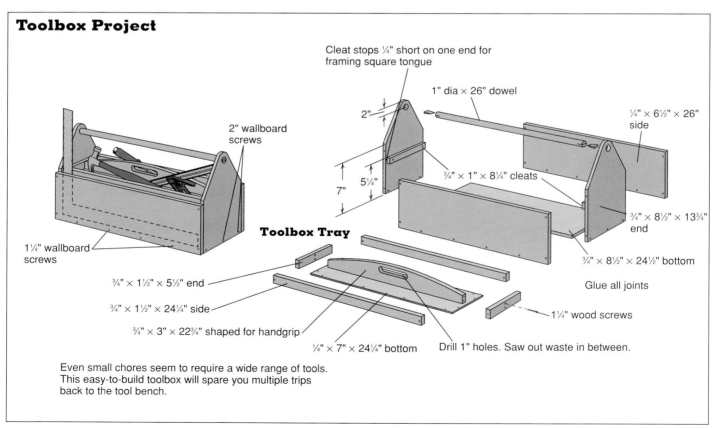

Cleat stops ¼" short on one end for framing square tongue

1" dia × 26" dowel

2"

5¼"

7"

¾" × 1" × 8¼" cleats

¼" × 6½" × 26" side

¾" × 8½" × 13¾" end

¾" × 8½" × 24½" bottom

Glue all joints

1¼" wood screws

2" wallboard screws

1¼" wallboard screws

Toolbox Tray

¾" × 1½" × 5½" end

¾" × 1½" × 24¼" side

¾" × 3" × 22¾" shaped for handgrip

¼" × 7" × 24¼" bottom

Drill 1" holes. Saw out waste in between.

Even small chores seem to require a wide range of tools. This easy-to-build toolbox will spare you multiple trips back to the tool bench.

WOODWORKING POWER TOOLS

As you gain experience and confidence, you'll want to expand your tool collection so you can tackle more ambitious projects. Power tools will help you expand your repertoire. This section explains the most common woodworking power tools and how to get the most from them.

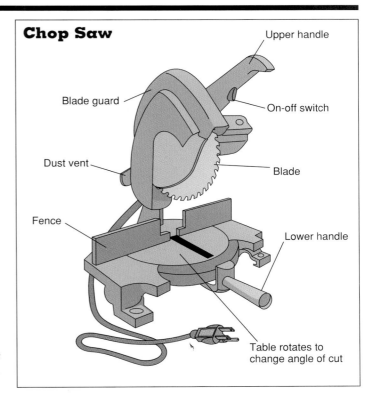

Chop Saw

- Upper handle
- Blade guard
- On-off switch
- Dust vent
- Blade
- Fence
- Lower handle
- Table rotates to change angle of cut

Chop Saw

The power miter box, commonly called a chop saw, is a great time-saver for all kinds of carpentry work. It cuts moldings and trim to length quickly and accurately, but its use is not limited to finish carpentry. A chop saw can make perfectly square cuts in 2×4 stock, making it handy for all kinds of framing work as well. With an auxiliary table, it's easy to handle long pieces of wood, and repeat cuts are a snap.

A chop saw is designed for crosscutting, so you should always use a saw blade designed for this task. A 40- to 60-tooth carbide blade with an alternate-top-bevel (ATB) or triple-chip grind gives the smoothest cuts.

Choosing a Chop Saw

Chop saws are available in sizes that handle blades from 8½ to 14 inches in diameter. The 10-inch size is the most common. All models have a retractable blade guard that moves back as the saw blade is lowered into the work. One of the most important things to look for when purchasing

a chop saw is a blade guard that works smoothly. A good guard won't bind up on the piece being cut and should lift out of the way easily to give you a good view of your cut-off mark.

Some chop saws have a fixed table with a wood insert; others have a metal table with a circular section in the middle that rotates with the saw head. Wood tables eventually get carved out in the middle from repeated miter cuts. If the table inserts aren't replaced frequently, short cutoffs can drop into the recess and get caught in the saw blade. Saws with rotating tables don't have this problem because the saw blade always drops into the same slot in the table.

One variant of the chop saw is the sliding compound miter saw. These saws are more expensive than conventional miter saws, but are more versatile. They can handle larger stock—up to 12 inches wide with some models—and can cut bevels, miters, and compound angles. Most models have an adjustable stop that limits the depth of the cut, allowing you to make dado cuts with repeated passes of the saw blade. A sliding compound miter saw can handle most of the cross-cutting chores of a radial

arm saw, with the added advantages of compact size, portability, and rapid angle adjustment.

Reciprocating Saw

This saw is the carpenter's equivalent of the surgeon's scalpel. It's a real help for re-modeling work because it can cut away parts of an existing structure with surgical precision, reaching into tight spots that no other tool can access. A reciprocating saw can quickly cut a new opening for a door or window, remove a door jamb, or cut holes for pipes and heating ducts. It also makes an excellent power hacksaw.

Reciprocating saws come in one-, two-, and variable-speed models. A variable-speed saw is the easiest to use, because it allows you to start a cut with

the blade moving very slowly, then speed it up once the blade is fully engaged in the material. A slower speed is also useful for cutting metal because it reduces the chances of over-heating and dulling the blade.

If you use a reciprocating saw to cut into a wall, remember that you may encounter hidden electrical wires and plumbing. Carefully remove a portion of the wall covering first to ascertain what lies beneath. Then disconnect electrical circuits and shut off water and gas supply lines before proceeding further.

Reciprocating Saw Blades

The secret of success when using a reciprocating saw lies in choosing the right blade for the job. Blades are available for cutting wood, plastic, metal, and nail-embedded

Reciprocating Saw

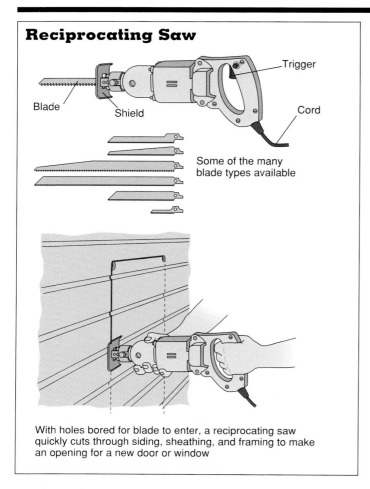

Blade
Shield
Trigger
Cord

Some of the many blade types available

With holes bored for blade to enter, a reciprocating saw quickly cuts through siding, sheathing, and framing to make an opening for a new door or window

Saber Saw

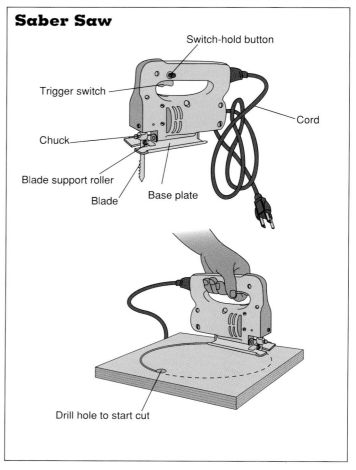

Switch-hold button
Trigger switch
Cord
Chuck
Blade support roller
Blade
Base plate

Drill hole to start cut

---

wood. Lengths vary from 3 to 18 inches, with 6- and 12-inch blades the most common types.

In general, you should use the shortest blade that will do the job. Long blades tend to whip around as they work, making them harder to control. Sometimes a long blade is necessary, as when removing shingles, reaching into a deep recess, or cutting very thick material. In these situations, try to start the cut slowly until the blade is adequately supported to run at a faster speed.

Bimetal blades are usually the best choice for remodeling work. They have a cutting edge made of very hard steel, welded to a back of softer, more flexible metal. This type of

blade remains sharp longer than other types, resists breakage, and survives an occasional encounter with a hidden nail.

There is no such thing as an all-purpose reciprocating saw blade. For rough-cutting wood, choose a stiff blade with 5 to 7 teeth per inch (TPI). For cutting nails or nail-embedded wood, use a thin bimetal blade with 14 to 16 TPI. To cut metal, use the same criteria as when selecting hacksaw blades (see page 27). If you need to cut flush to an adjoining surface, a long blade allows you to hold the saw at an angle and bend the blade slightly while you make the cut.

Saber Saws

The saber saw is a very useful tool, and recent design advances have made it easier and more convenient to use. Saber saws make curved cuts better than any other portable power tool. They also excel at making cuts in the middle of a board, as when cutting out for electrical outlets in paneling or notching deck boards around posts.

Saber saws come with one-speed, two-speed, three-speed, and variable-speed motors. Try to select a saw with at least two speeds. Use the faster speed range for soft materials like wood, and the slower speeds for hard mater-

ials like steel. A slow speed also helps if you are trying to make very accurate cuts freehand in wood.

When you buy a saber saw, look for one that has a blade support roller. This is a small grooved disk located just above the saw table and behind the blade. It supports the back of the blade as it moves up and down, providing a squarer cut and reducing blade breakage considerably.

Some saber saws come with an orbital action feature. With these saws, the blade moves up and down in an elliptical path: forward on the up (cutting) stroke and back on the down stroke. This motion clears sawdust from

Saber Saw Blades

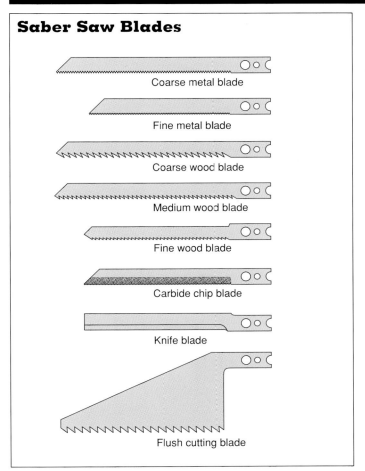

Coarse metal blade

Fine metal blade

Coarse wood blade

Medium wood blade

Fine wood blade

Carbide chip blade

Knife blade

Flush cutting blade

Plunge Router

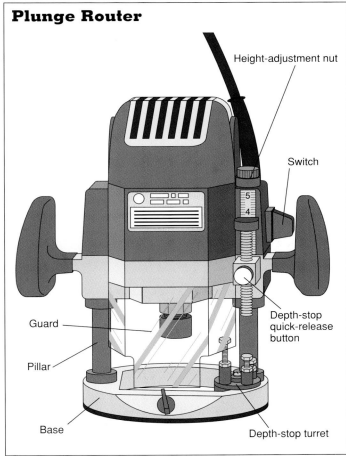

Height-adjustment nut

Switch

Depth-stop
quick-release
button

Guard

Pillar

Base

Depth-stop turret

---

the kerf, resulting in smoother, faster cutting. The easiest way to make a cutout into a board with a saber saw is to drill a hole first, then lower the blade of the saw into the hole and complete the cut.

Saber Saw Blades

Blades for these saws are available for cutting all kinds of materials, from rubber to wood to stainless steel. The maximum depth of cut for most models is about 2½ inches. Use the coarsest possible blade if you are cutting in thick wood because fine-toothed blades tend to clog up with sawdust.

Always use a narrow blade if you have to cut very tight circles or curves so the kerf does not restrict the turning ability of the tool. Never force a blade around a corner or it may snap. If necessary, back up and make several passes through the waste portion to complete the curve.

Routers

Once you use a router, you'll wonder how you ever got along without one. With the large choice of bits available, you can cut decorative molded edges, make dado and groove cuts, and cut mortises for door hinges. A router can trim thin

laminates with ease, make decorative dovetail joints, and cut perfect circles of any radius.

The router consists of a motor that drives a shaping bit at tremendous speeds—up to 27,000 rpm, compared with 2,200 for an average drill. The motor can be raised and lowered to adjust the depth of the cut. A scale on the housing allows you to read the amount of adjustment.

Choosing a Router

Routers come in two basic configurations. Fixed-base routers have a motor that fits into a sleeve that is part of the base of the tool. Depth adjust-

ments are made by sliding the motor up or down in the sleeve and clamping it in position with a nut on the housing. Plunge routers have a motor that rides on a pair of spring-loaded posts attached to the base. This arrangement allows the router to be lowered into the work with the motor running, a very useful feature for mortising and some other routing jobs.

Routers range in horse-power (hp) from ½ to 3 hp. The smaller models are suitable only for light cutting, such as trimming laminates. Since you will probably use a router more than you think, choose at least a 1 hp model.

Using a Router

A router is guided through the cut with a pilot bearing on the tip of the bit that rides along the edge being shaped, or a fence attached to the router or the workpiece, or a guide bushing that encloses part of the bit and rides along a cutout in a specially shaped template.

Even though a router turns at very high speed, there is a limit to how much material it can remove in one pass. That limit depends on the material being cut, the bit type, and the power of the router. If the tool seems to be slowing down excessively or makes funny noises during the cut, it's best to raise the cutter and take two or more passes.

When using a router, always be aware of the rotation of the bit and how it affects the cut being made. (The bit always turns clockwise as you look down on the top of the tool.) If you're routing the edge of a board, feed the tool into the work so that the cutting edges of the bit are moving the same direction as the feed. If you try to make the cut backward, the bit will hog into the material and the tool will jump out of control. If you're using a fence, make sure the cutting action of the bit will pull the tool snugly up to the fence; if you go the wrong way, the bit is sure to wander and ruin your work.

One problem often encountered by beginners is burned spots on routed edges. These are caused by allowing the spinning bit to stay too long in one spot. Router burns can be very difficult to remove, even with heavy sand-ing. Always keep the router moving through the cut, and move the bit away from the work when you stop.

Router Bits and Accessories

Cutting bits for routers are made of carbon steel or tungsten carbide. Carbide bits outlast steel many times over, but are more expensive. The cost of a set of router bits can easily exceed the cost of the tool itself, so it is best to start with a few basic cutters and buy new ones as you need them.

Straight Cutters

These bits are used to make grooves, dadoes, and mortise cuts. They require a fence, edge guide, or template-and-bushing setup to guide the cut.

Shaping Cutters

These cutters shape moldings, make decorative edges on boards, and cut rabbets and grooves. They are guided by a ball bearing or steel nub that rides along the edge of the board.

Sanders

Nobody enjoys sanding, but it is the essential final step in preparing wood for finishing. You can do all your sanding with sheets of sandpaper and elbow grease, but hand-sanding quickly becomes a tedious and time-consuming chore. Several types of power sanders are available to speed up the job.

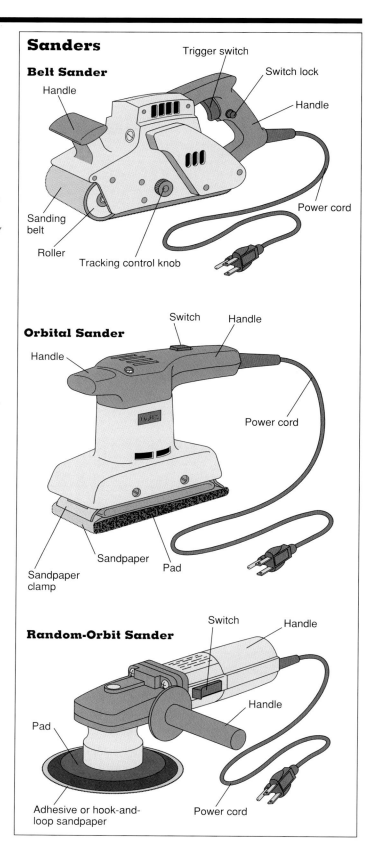

Sanders

Belt Sander
Handle
Trigger switch
Switch lock
Handle
Sanding belt
Roller
Tracking control knob
Power cord

Orbital Sander
Handle
Handle
Switch
Power cord
Sandpaper clamp
Sandpaper
Pad

Random-Orbit Sander
Switch
Handle
Handle
Pad
Adhesive or hook-and-loop sandpaper
Power cord

Belt Sanders

No sander works as quickly as the belt sander. It can also ruin a piece of work if used carelessly, so practice your technique on scrap wood.

When using a belt sander, you can avoid gouging the wood by keeping the base of the sander flat on the work and moving it the full length of the piece in long, slightly overlapping strokes. Always keep the sander moving. At the ends of the board, allow about one third of the base plate of the tool to extend beyond the edge before changing directions.

Always sand with progressively finer grits of sandpaper. If the piece is very rough to start with, flatten it with a coarse 60-grit belt, holding the sander at a 45-degree angle to the grain for faster cutting. Follow up with 80- and 120-grit belts, working only with the grain and taking care to sand out all scratches from the previous belt before changing grits. Finish with an orbital sander or by hand-sanding with finer grits of paper.

Orbital Sanders

The orbital sander, also called a finish sander, works more slowly but produces finer results than the belt sander. The orbital sander moves in a tight circle, each stroke covering only about ¼ inch. Because it is moving in all directions, you don't have to follow the grain.

Random-Orbit Sanders

These sanders use a circular sanding disk that simultaneously spins and oscillates back and forth. This produces a sanding action that is nearly as fast as a belt sander, yet as smooth as an orbital sander.

Special sanding disks are attached to the sander with pressure-sensitive adhesive (PSA) or hook-and-loop fastening. PSA disks are the least expensive, but must be discarded when changing grits. Hook-and-loop disks can be removed and reused until they are worn out.

Sanding and Wood Dust

All machine woodworking tools produce dust, but sanders produce very fine particles that take a long time to settle out of the air. Some individuals find that sanding dust causes allergic reactions and respiratory irritation. All woodworkers should take steps to protect themselves from exposure to airborne wood dust.

Most power sanders made today come from the factory with some sort of dust collection bag as part of the tool. These bags should be cleaned frequently and replaced if they become torn or damaged. Even the best bags will not catch all of the dust, so you should also wear a snug-fitting particle mask when using a power sander.

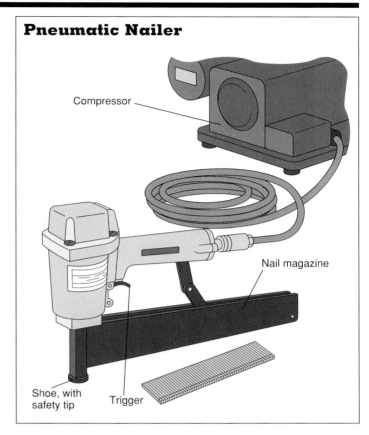

Pneumatic Nailer

Compressor

Nail magazine

Shoe, with safety tip

Trigger

Other Useful Power Tools

Besides the power tools shown in the previous pages, many others are available to the home woodworker. Because of their cost and specialized nature, they aren't candidates for the essential toolbox, but are well worth considering for extensive projects.

Pneumatic Nailers

Air-powered nail guns have been around since the 1930s, but in recent years have nearly replaced the hammer for professional carpenters and cabinetmakers. The cost of a nail gun, along with the air compressor and hoses needed to drive it, makes owning a pneumatic nailer impractical for all but the most fanatical do-it-yourselfer. But if you choose to rent one for a large job, its advantages will become apparent immediately.

Speed is only the most obvious advantage of a nail gun. Because it requires just one hand to operate, you have the other hand free to hold pieces in place for fastening. And because a gun injects the nail into the wood, your work absorbs a lot less pounding than with hand nailing. That translates into fewer splits and dislocated backing. Best of all, you don't have to spend years perfecting your hammering technique—and don't have to worry about smashed fingers.

For framing projects, you should use a gun designed to drive full-headed framing nails up to 16 penny. These guns really shine when nailing plywood wall and roof sheathing because they can drive thousands of nails in a fraction of the time it would take to do the job by hand. Framing guns are designed to countersink the nail head slightly below the surface of the wood, but most models have adjustments or attachments that allow the head to be set flush with the surface when required.

Another type of nail gun is designed for finish carpentry work. Finishing nailers come in two basic versions. One drives a full-sized finishing nail, usually 1½ to 2½ inches long. These guns are excellent for hanging door jambs, assembling cabinets, and installing moldings and trim. The other type, known as a brad nailer, drives light-gauge brads, usually ⅝ to 1½ inches long. Brad nailers are great for installing small moldings, and their heads are so small that they often don't need puttying. Some brad nailers are electric powered, so they don't require an air compressor.

Nail guns can be dangerous if improperly used. Sometimes a nail follows the grain of the wood and the point pops out in an unexpected place. Always keep your hands well away from the nose of the gun. If a nail hits another fastener or some other obstruction, it can ricochet, so protect your eyes with safety glasses whenever you use a nail gun.

Power Planes

These tools look somewhat like hand planes, but they have a cylindrical cutter head driven at high speed by a motor. Power planes are handy for straightening crooked studs prior to applying wall finishes, for knocking off high spots on twisted header beams, and for flattening twisted or bowed lumber. A power plane quickly removes saw marks from the cut edges of boards, leaving a smooth surface that requires very little sanding.

One useful accessory for the power plane is an adjustable fence that allows you to plane accurate square or beveled edges. This feature is good for edge-joining boards or trimming doors to fit their openings.

Biscuit Joiners

This tool, also known as a plate joiner, has a small circular saw blade that can be plunged into the work, leaving a shallow kerf. After matching cuts are made in two pieces to be joined, a football-shaped biscuit of compressed wood is glued into the saw kerfs to bridge the joint. Moisture in the glue causes the biscuit to swell, filling the cut and making a very strong joint.

A biscuit joint is as strong as one made with dowels, and it's much easier to make. The shape of the saw kerf allows some sideways adjustment during gluing, so perfect alignment is not necessary. Biscuit joiners can make sturdy, durable joints in solid wood, plywood, or composition boards.

Power Plane

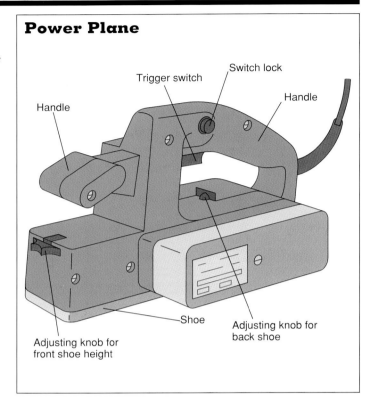

Handle
Trigger switch
Switch lock
Handle
Adjusting knob for front shoe height
Shoe
Adjusting knob for back shoe

Biscuit Joiner

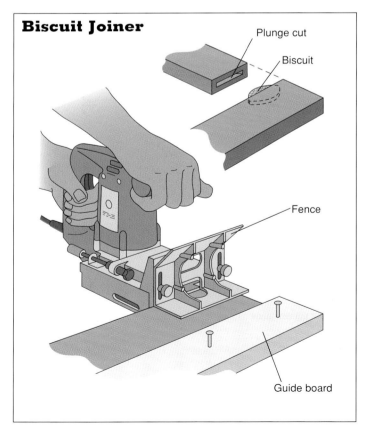

Plunge cut
Biscuit
Fence
Guide board

Table Saw

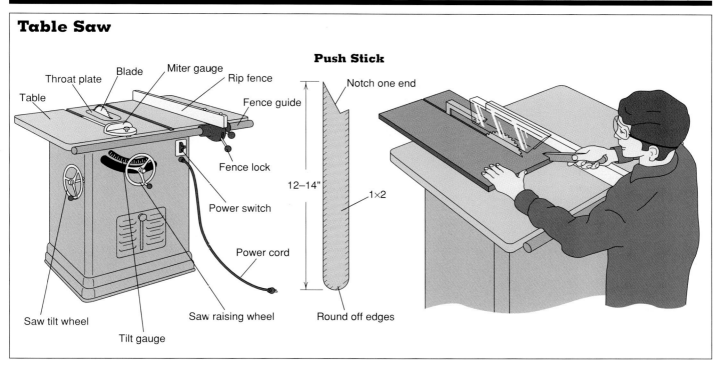

Push Stick

Notch one end

12–14"

1×2

Round off edges

Table · Throat plate · Blade · Miter gauge · Rip fence · Fence guide · Fence lock · Power switch · Power cord · Saw tilt wheel · Tilt gauge · Saw raising wheel

Stationary Power Tools

If you get seriously infected with the woodworking bug or face a major remodeling project, you'll want to invest in stationary power tools like a table saw or radial arm saw. These tools will allow you to do quick stock preparation that would be impractical with hand tools or portable power tools.

Table Saw

This saw occupies center stage in any wood shop. Its primary function is to rip stock to width, but with the addition of a few accessories and shop-made jigs, it also does accurate crosscuts and a wide variety of cabinet joints, including rabbets, dadoes, and grooves.

The blade on a table saw can be adjusted up and down for cutting different thicknesses of material, and it tilts to one side for bevel cuts of 45 to 90 degrees.

Choosing a Table Saw

Large table saws are heavy and should be considered a permanent shop tool, but several small portable table saws are available. Some have enough horsepower to cut stock up to 2 inches thick with ease. Portable saws are smaller and require two people to handle large pieces of wood. If you work alone, it's a good idea to add extension tables to the saw.

Table saws are designated by the size of the largest saw blade they can handle. A 9- or 10-inch saw is adequate for a home shop. The most important feature to look for in a table saw is a good fence. It is surprising how many manu-

Dado Blades

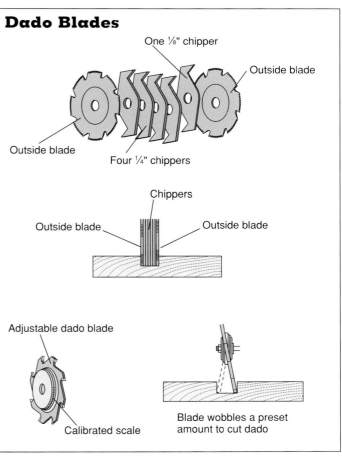

One ⅛" chipper

Outside blade

Outside blade

Four ¼" chippers

Chippers

Outside blade

Outside blade

Adjustable dado blade

Calibrated scale

Blade wobbles a preset amount to cut dado

Dado Cuts

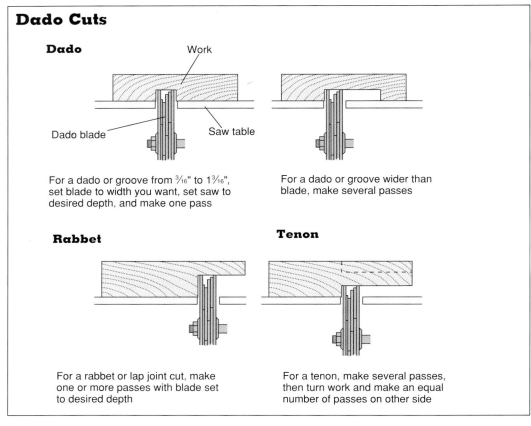

Dado

Work

Dado blade

Saw table

For a dado or groove from ³⁄₁₆" to 1³⁄₁₆", set blade to width you want, set saw to desired depth, and make one pass

For a dado or groove wider than blade, make several passes

Rabbet

For a rabbet or lap joint cut, make one or more passes with blade set to desired depth

Tenon

For a tenon, make several passes, then turn work and make an equal number of passes on other side

Radial Arm Saw

Miter scale

Swivel-latch knob

Power switch

Radial arm

Rip scale and carriage-lock knob on right side of arm

Blade guard

Saw blade

Arm-latch knob

Pull handle

Fence

Bevel-lock scale and knob

Table

Antikickback and spreader assembly

Elevation crank

facturers scrimp on this essential part of the tool. A good fence should move easily to any position on the saw table without binding, it should lock into position securely, and it should always remain exactly parallel to the saw blade. If you have a saw with a balky fence, you can purchase an aftermarket fence that will improve the saw's performance dramatically.

Table Saw Blades

You will get much better service from a table saw if you replace the steel blade that comes with the machine with a carbide-tipped blade. A carbide blade produces smoother cuts and lasts much longer. For all-around use, choose a combination-tooth blade that is

designed for crosscutting and ripping. If most of your work involves ripping boards to width, a thin-kerf rip blade cuts faster and places less strain on the saw motor. For cutting sheets of plywood, laminates, and composition boards, a fine-toothed triple-chip blade or a blade with alternate-top-bevel teeth (ABT) gives the smoothest cuts.

A dado blade allows you to cut rabbets, grooves, and dadoes. These blades come in two different types. The best dado cutters consist of a pair of saw blades and a set of chipper blades of various thicknesses that can be stacked together to produce grooves from ¼ to 1³⁄₁₆ inches in width (wider grooves can be made with multiple passes). The other type, called an adjustable dado, is designed to wobble as it spins. The amount of wobble can be adjusted to vary the width of the groove.

Using a Table Saw

Before making any cut on a table saw, first set the blade to the correct height. The blade should protrude above the stock being cut by the height of one saw tooth.

To make the basic rip cut, first check to make sure the fence is parallel to the blade by measuring the distance between the fence and the front and rear edge of the blade. Push the stock into the saw. If the piece tends to drift away from the fence or if one of the cut surfaces appears to be burned, you should recheck the fence and adjust it if necessary.

Whenever you rip narrow pieces of wood, use a push stick to make the last part of the cut. This will keep your fingers safely away from the spinning blade. Make a push stick by notching one end of a piece of 1×2 scrap, and round off the handle so it fits comfortably in the palm of your hand.

Crosscuts are made with the miter gauge fixture that comes with the saw. The miter gauge runs in grooves machined into the saw table parallel to the blade. Before you use a miter gauge, it's a good idea to screw a piece of 1-by lumber to its face to give more support to the stock. For square cuts, use a framing square to set the gauge perpendicular to the grooves in the saw table. Use the protractor scale on the miter gauge to make miters and other angle cuts. Most miter gauges have built-in stops that can be adjusted to make accurate cuts at 45 and 90 degrees, the most-often used settings.

Radial Arm Saw

Unlike the table saw, on which the blade is mounted in a fixed table, the radial arm saw has a blade mounted to a movable motor that slides on an extended arm. Radial arm saws are designed primarily for crosscutting, but it is possible to rotate the cutting head to make rip cuts as well.

To function properly, radial arm saws must be adjusted in several ways. The saw blade must be perpendicular to the saw table and parallel to the radial arm. The table must be parallel to the radial arm. The arm must be perpendicular to the fence on the table. The blade must be set to the correct height in relation to the saw table. Given that these adjustments are time-consuming and difficult to make and must be redone every time the angle settings are changed, the radial arm saw is less convenient than the table saw for most people. It's still handy, though, for quickly crosscutting lumber on framing jobs.

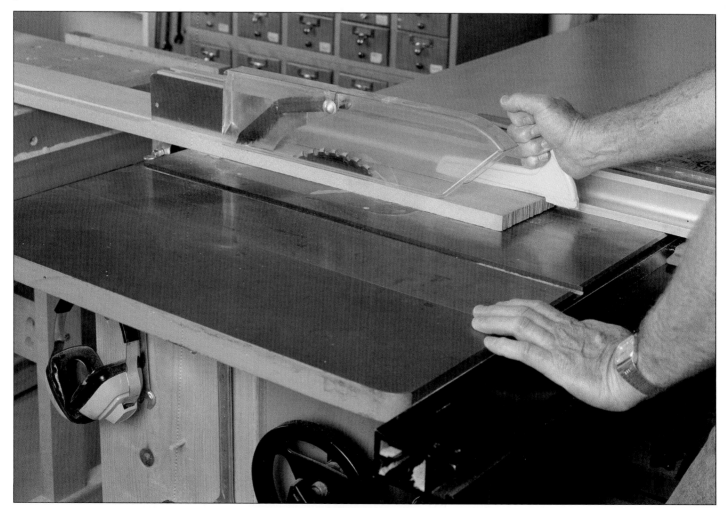

Before operating any power tool, be sure that the safety guards are attached properly and that you are familiar with safe operating techniques. This table saw has a blade guard and a splitter. The outfeed table behind the saw and a push block also ensure safe operation, as do safety glasses and ear protection for the operator.

CARPENTRY TECHNIQUES

You've familiarized yourself with all the materials carpenters use, you've assembled a tool kit, your plans are in place, and now you're ready to get to work. The following pages explain the basic steps required to build a simple structure from foundation to roof shingles. The emphasis, though, is on construction techniques, rather than on how to build any particular plan. Study the whole process before you start, because every step along the way depends on the work that has gone before. Armed with this knowledge, you'll be able to build anything from a garden shed to a small house—and tackle typical problems encountered in remodeling work.

Although this home is large and complex, the basic carpentry techniques for framing and finishing it are the same as for smaller buildings. After the foundation was formed and poured, a floor system consisting of mudsills, girders, floor joists, and a plywood subfloor was built over it. Then the walls were framed and tilted up, sheathing was added, and the roof was framed. The next step, after framing, will be to finish the roof.

FOUNDATIONS

No matter what kind of structure you plan to build, it will need a foundation to anchor it securely to the ground. Four types of foundations are covered in this section: post-and-pier, slab-on-grade, formed concrete, and concrete-block. The one you choose will depend on your budget and the type of structure you'll be building.

Foundation Layout

Before you begin to construct a foundation, you'll have to lay out the site. Regardless of the type of foundation you will be using, you should put up a set of batter boards and string lines to locate the corners of the building and establish the height of the floor. The batter boards are set 2 or 3 feet out from the building outline; this allows you to locate the corners of the building, remove the string lines to dig the footings, and replace them afterward to set forms or posts.

First determine the approximate locations of the building corners and mark them with stakes. You don't have to be precise at this stage—just try to get within 3 or 4 inches of the final dimensions. Tie string lines between the stakes and use a framing square to make the corners roughly square, so you can set batter boards.

Batter Boards

Batter boards are made from 1×3 or 1×4 stakes connected by 4-foot lengths of 1×4. Start by roughly measuring the footprint of your structure and driving the three stakes outside each corner. Use a hydro level, line level, or builder's level to mark each stake at the same elevation. For post-and-pier foundations, the marks should be level with the bottom of the floor girders. For concrete-block foundations, the marks should be even with the top of the blocks. For a slab floor, they should be at the same height as the top of the slab.

After marking post heights, use wallboard screws to attach the crosspieces to the stakes. Attach them a little higher on the stakes than needed, then tap the stakes down until they are all at exactly the right elevation, rechecking with the level as you go. Then brace the batter boards so they will not move when you pull the strings across them.

Stretch string (nylon mason's twine works best) across the batter boards so the strings intersect over the temporary corner stakes. Adjust them back and forth until they are at the right dimensions, parallel, and square. Make shallow saw cuts in the crosspieces at the string locations so you can remove and easily replace the string as needed.

Playhouse Plan

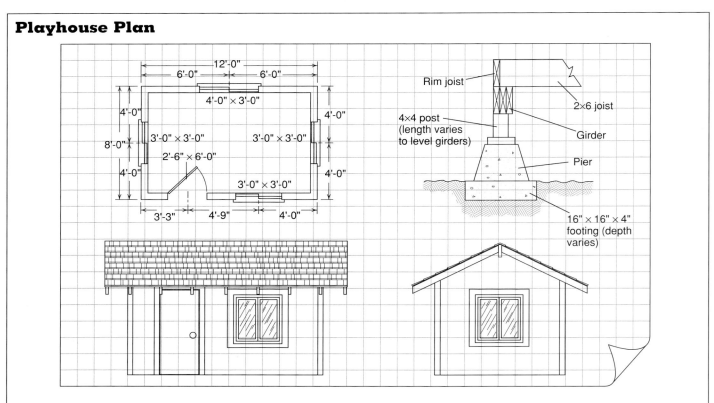

Laying Out the Building Site

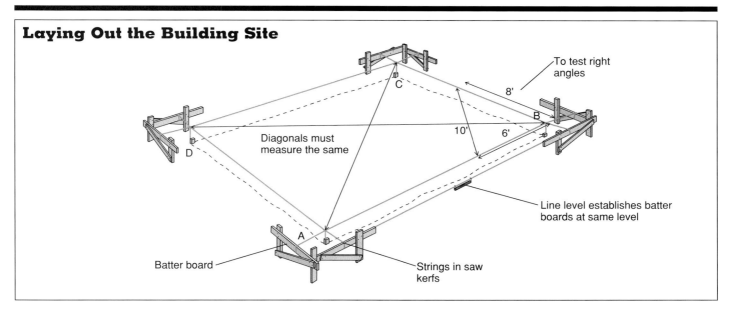

To test right angles

8'

10'

6'

Diagonals must measure the same

Line level establishes batter boards at same level

Batter board

Strings in saw kerfs

Playhouse Framing

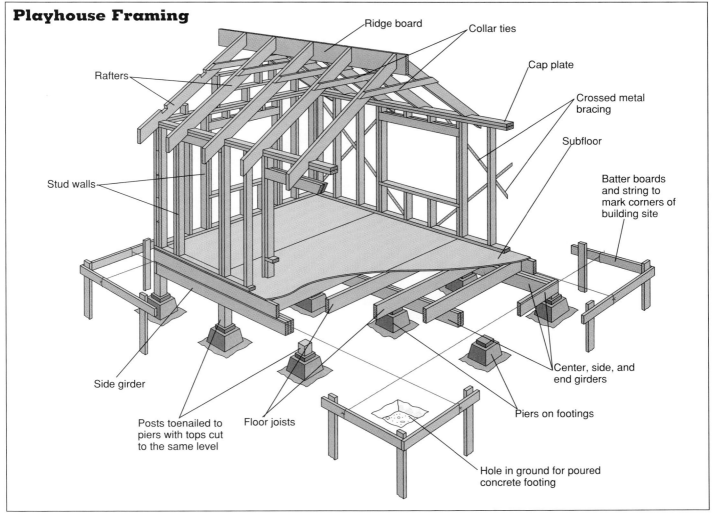

Ridge board

Collar ties

Cap plate

Crossed metal bracing

Subfloor

Batter boards and string to mark corners of building site

Rafters

Stud walls

Side girder

Center, side, and end girders

Piers on footings

Posts toenailed to piers with tops cut to the same level

Floor joists

Hole in ground for poured concrete footing

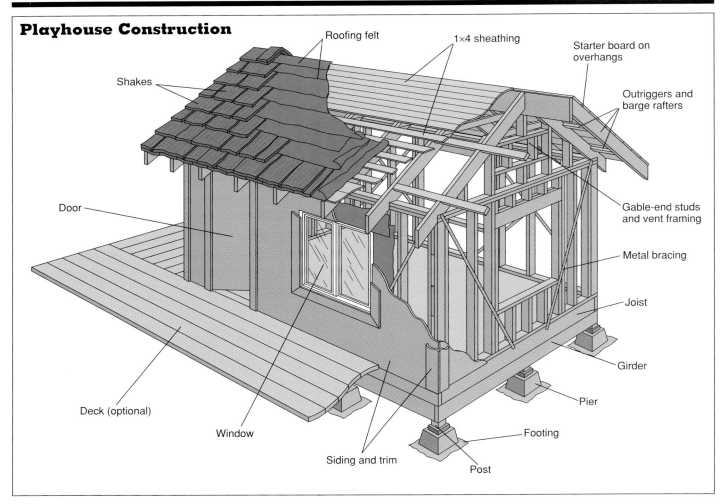

Playhouse Construction

Roofing felt

1×4 sheathing

Starter board on overhangs

Shakes

Outriggers and barge rafters

Door

Gable-end studs and vent framing

Metal bracing

Joist

Girder

Pier

Deck (optional)

Footing

Window

Siding and trim

Post

Squaring Up

It's essential that the foundation be built as square as possible. If it isn't, you will be making corrections all through the construction process, right up to the top of the roof. There are two basic methods for squaring up a foundation layout: measuring diagonals and the 3-4-5 triangle method.

If your building is a simple rectangle, the easiest way to square it is to measure the diagonal distances between opposite corners. If the dimensions are equal, the building will be square. This method works only if opposite sides are parallel, so every time you

adjust one string you will have to adjust at least one other and recheck your measurements. Have an assistant hold one end of the tape while measuring.

For buildings that are odd-shaped or contain offsets, start your layout by squaring up two of the main walls with the 3-4-5 triangle method and then use these lines as a reference for laying out the rest of the building. The 3-4-5 method is based on the fact that one corner of a triangle with sides measuring 3 feet, 4 feet, and 5 feet (or any multiples thereof) will always be exactly 90 degrees. Thus, triangles of 6, 8, and 10 feet; 9, 12, and 15 feet,

or 30, 40, and 50 feet will all give you a true right angle. For greatest accuracy, use the largest possible triangle. Mark the measurements on the layout strings with pins or short lengths of twine knotted around the string.

Post-and-Pier Foundation

The post-and-pier foundation is the easiest and least expensive foundation to build and is well suited to small outbuildings and deck construction. This foundation may need footings under each pier if it is built on a slope or if local

codes require. However, footings are not always necessary.

Piers are precast tapered concrete blocks that are 12 inches square at the bottom and have a 2×6 block of redwood or pressure-treated wood set in the top. Piers are placed in position on the ground or on footings to support the posts, which in turn form a level base for attaching the girders.

Installing Piers and Posts

If you are building on sloping ground, it is advisable to dig footing holes 6 to 12 inches

Precast Pier

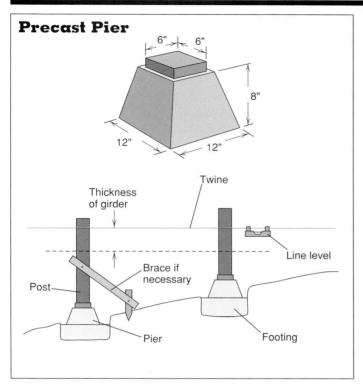

Girders

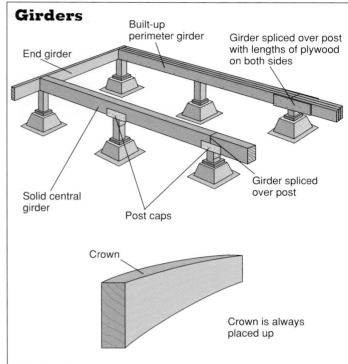

deep and fill them with concrete. Set the piers in place while the concrete is still wet to anchor them securely. Check the pier top for level in all directions.

The number of piers you will need and the distance between them are determined by the size of the girders and the floor joists. See the span tables on page 93 to determine the allowable spacing between supports for floor joists.

Once your layout is squared, use a plumb bob to set the piers so the posts will be centered on them. For 4×4 posts, the edges of the wood blocks will be 1 inch outside the layout strings. Set the corner piers first, then the intermediate ones. All the piers do not have to be on the same level. You can compensate for differences in elevation when you cut the posts to length.

If you set batter boards to the height of the bottom of the girder, finding the lengths of the posts is easy—just measure down from the strings to the tops of the piers. To cut the center posts, first install the perimeter posts and pull string lines between them to determine the required length. Nail each post to its pier with four 8-penny (8d) galvanized nails.

On fairly level ground, posts may range in height from a few inches to a foot or two. If they are higher than that, brace them with 1×4s. When all the posts are in place, install the floor girders.

Installing Girders

Girders span the posts and form the building perimeter. They run down the long sides of the building. If the span between them is wide, one or more intermediate girders

may be required. Girders are usually made of 4-by lumber or built up from two or three thicknesses of 2-by material. The ends of the girders are tied together across the width of the building by single 2-by joists the same depth as the girders (single joists are used here because they are non-load-bearing).

Cut the girders to length, making them 3 inches shorter than the length of the building to allow for the two end joists. Make them as long as possible, but if the building is longer than 20 feet, you'll have to splice them in the middle. Make splices over a post and tie the two ends together with strips of plywood or metal straps. If you are using built-up girders, assemble them with pairs of 16d nails spaced 12 inches apart.

Install the girders by toe-nailing them to the tops of the

posts. For a stronger connection, use sheet-metal post caps. Nail the two end joists to the ends of the girders, and then recheck the whole assembly for square, making sure the framework is within ½ inch of being perfectly square. With the girder assembly completed, you are ready to frame the floor (see page 62).

Slab-on-Grade Foundation

A concrete slab foundation rests directly on the ground. It's an excellent choice for a small garage, shop, or barn because having the floor close to the ground makes it easier to move things in and out.

For small outbuildings and overhead structures, perimeter footings are not necessary. Permanent structures, such as garages and room additions, should have a continuous footing around the edge of the slab, as well as footings in the middle for any bearing posts or walls located within the building footprint. (You'll also need to install plumbing lines and electrical conduit before the slab is poured.) Most building codes require that slab footings be located below the frost line; however, this is not always practical for small structures. Check with the local building department for footing requirements.

Trenches, Footings, and Forms

You will need a level, well-drained site for the slab. If you will be including footings, dig the trenches 3 or 4 inches out-side the building line to allow room for forms and stakes. Set the perimeter forms level and square, aligning the top inside edge of the form boards with the batter-board strings. Brace the forms to the outside with stakes every 4 feet or so.

Use the forms as a guide to grade off the inside area to a depth 8 inches below the top of the forms. Put down a 2-inch layer of sand, followed by a vapor barrier of 6 mil plastic to prevent the slab from wick-ing up moisture. Cover the vapor barrier with another 2-inch layer of sand. The tops of the forms should now be 4 inches above the sand base, which will give you a 4-inch slab, the standard thickness for light construction.

Install the footing reinforc-ing bars, bending them around

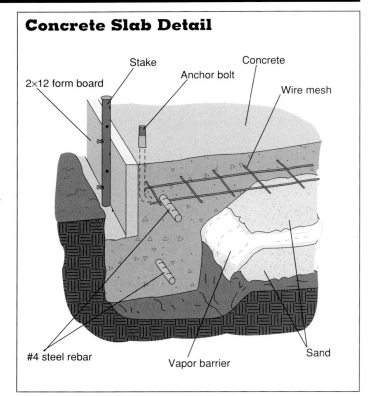

Concrete Slab Detail

Stake
Concrete
2×12 form board
Anchor bolt
Wire mesh
#4 steel rebar
Vapor barrier
Sand

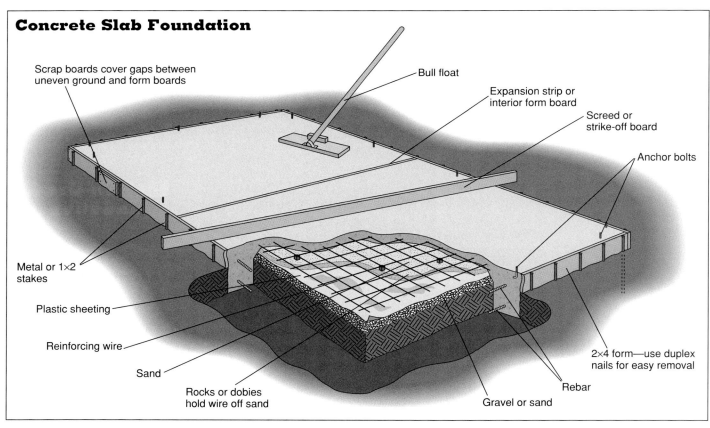

Concrete Slab Foundation

Scrap boards cover gaps between uneven ground and form boards
Bull float
Expansion strip or interior form board
Screed or strike-off board
Anchor bolts
Metal or 1×2 stakes
Plastic sheeting
Reinforcing wire
Sand
Rocks or dobies hold wire off sand
Gravel or sand
Rebar
2×4 form—use duplex nails for easy removal

Perimeter Footing

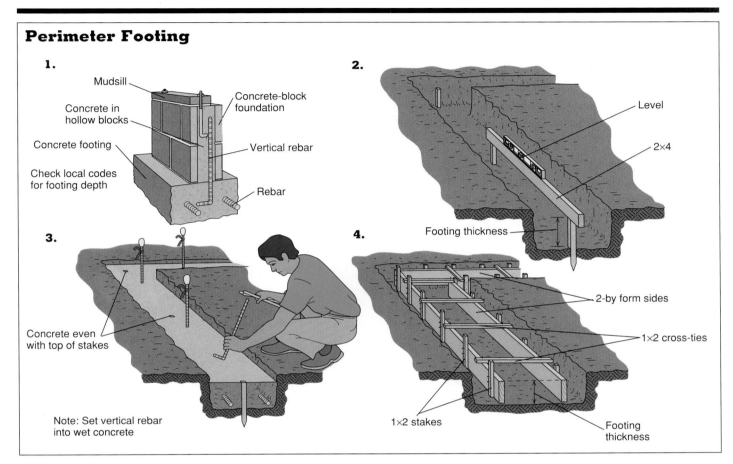

1.
- Mudsill
- Concrete in hollow blocks
- Concrete footing
- Check local codes for footing depth
- Concrete-block foundation
- Vertical rebar
- Rebar

2.
- Level
- 2×4
- Footing thickness

3.
- Concrete even with top of stakes
- Note: Set vertical rebar into wet concrete

4.
- 2-by form sides
- 1×2 cross-ties
- 1×2 stakes
- Footing thickness

the corners. Overlap splices at least 18 inches and tie them together with wire. Roll out 6-inch wire reinforcing mesh over the sand.

If you are pouring a 4-inch slab without footings, just grade the site level and form the perimeter with 2×4s. You can skip the trenches, reinforcing bars, sand, and vapor barrier, but be sure to include the reinforcing mesh.

Just before you pour, hose everything down with water so the slab won't dry out prematurely.

Pouring a Slab

Unless your slab is very small (less than ½ cubic yard), buy concrete from a concrete company and have it delivered

in a ready-mix truck. If the truck can't back up to the site, you will have to move the concrete with wheelbarrows or hire a concrete pumper. Pumpers are worth the extra money for all but the smallest jobs. Most pumping outfits will even order the concrete for you and arrange delivery. You should have payment ready for the pumper and the ready-mix company when the trucks arrive on the job.

Fill the forms to the top, pulling the reinforcing mesh up into the wet concrete as you go. Level the concrete off by dragging a 2×4 (known as a screed or strike-off board) across the forms with a saw-

ing motion. Smooth the surface immediately with a bull float, a tool that will work the gravel down below the surface and bring the finer particles to the top.

Install anchor bolts while you wait for the concrete to set up enough to trowel. Bolts should be placed within 12 inches of corners and door openings and no more than 6 feet apart. Set them about 2 inches in from the form boards and let them stick 2½ inches above the top of the slab.

When finishing concrete, timing is everything. Keep testing the surface with a trowel—as soon as it stops bringing water to the surface, it's time to get to work. Use long sweeping strokes, keep-

ing the leading edge of the trowel slightly elevated so it won't dig in. As the concrete hardens, raise the edge higher.

For a smooth finish, go over the slab twice. If you want a nonskid surface, drag a push broom over the slab immediately after the second troweling.

Formed Concrete Foundation

This is the most common type of foundation for a raised-floor building. The foundation is shaped like an inverted T, with the footing and foundation wall poured at the same time.

Pouring a Slab

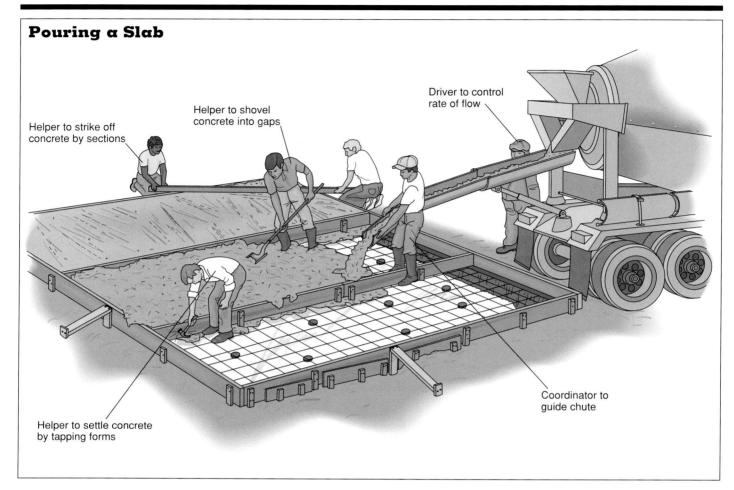

Helper to strike off concrete by sections

Helper to shovel concrete into gaps

Driver to control rate of flow

Coordinator to guide chute

Helper to settle concrete by tapping forms

Building the Forms

The forms for a concrete foundation wall are usually constructed of 2-by lumber. (You can save money if you make floor joists from the form material, but expect about 25 percent of the stock to be unusable.)

To set the forms, first use a plumb bob to position the outer row of stakes 1½ inches outside the batter-board string lines. Drive them in place 4 to 6 feet apart. Then cut a piece of scrap wood 3 inches longer than the width of the foundation wall and use it as a spacer to set a stake directly opposite each outer stake.

Install the topmost outer form board first. Set it a little high, nail it to a stake at each end, then carefully tap the stakes down until the top inner edge lines up with the string line. Brace it in place from the outside with 2-by diagonal supports. Fill in the rest of the outer form boards to the top of the footing. Space them ⅛ inch apart (use a 16d nail as a spacer) so the metal form ties can be added later.

Place the inner form boards. Then install the reinforcing bars for the footing, bending them around the corners. Overlap the bars by at least 18 inches and splice them together with wire.

Finishing a Slab

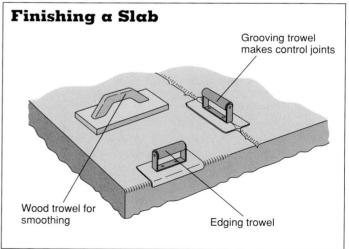

Grooving trowel makes control joints

Wood trowel for smoothing

Edging trowel

The two sides of the forms must be tied together so they won't spread apart when the pour begins. Metal form ties, available for 6-, 8-, and 12- inch-thick foundation walls, are an inexpensive and reliable way to hold the forms together. Slip the ties into the gaps between boards, then

Forming a Foundation Wall

Pouring Concrete

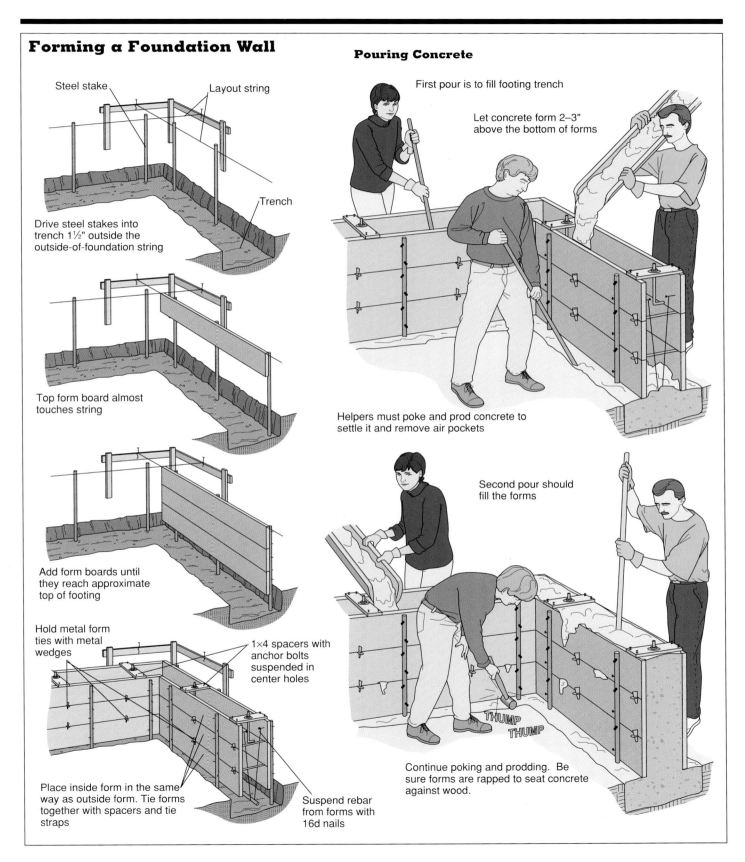

Steel stake

Layout string

Trench

Drive steel stakes into trench 1½" outside the outside-of-foundation string

Top form board almost touches string

Add form boards until they reach approximate top of footing

Hold metal form ties with metal wedges

1×4 spacers with anchor bolts suspended in center holes

Place inside form in the same way as outside form. Tie forms together with spacers and tie straps

Suspend rebar from forms with 16d nails

First pour is to fill footing trench

Let concrete form 2–3" above the bottom of forms

Helpers must poke and prod concrete to settle it and remove air pockets

Second pour should fill the forms

THUMP THUMP

Continue poking and prodding. Be sure forms are rapped to seat concrete against wood.

insert metal wedges into the slots outside of the forms. Once all the ties are in place, the rebar at the top of the wall can be installed and wired to the form ties.

Pouring the Foundation

The foundation wall should be poured in two stages. The first pour will fill the footing trench to the bottom of the form boards. By the time you get back around to the starting point, this concrete should be set up enough to pour the stem wall up to the top of the forms. Work the concrete with a stake or shovel handle to eliminate air pockets in the concrete, and strike it off level with a wood float.

When the concrete starts to set up, refloat the wall with a wood float. Add concrete to the top of the forms wherever it has settled. Once the concrete is firm, but not completely hardened, remove all the stakes. Also check to make sure that none of the lower form boards is embedded in the concrete.

The form boards can usually be removed the next day. Clean them with a scraper, a stiff brush, and lots of water, and stack them carefully for reuse. Remove the protruding ends of the form ties by twisting them off with a pair of pliers.

Concrete-Block Foundation

The concrete-block foundation requires footings, which are poured in a trench and on which the foundation wall rests. This type of foundation is the most difficult of the three types to construct, but it provides the best support for a raised wood floor.

Footing Trenches and Forms

The first step in building a block foundation is to dig the footing trenches. The rule of thumb for digging the footings is to put them on solid earth (not fill dirt) and below the frost line. In a cold climate, footings above the frost line will be heaved by the freezing earth, opening up cracks in the building and causing doors and windows to stick. Sites with expansive soils can also be subject to foundation heaving. Check with the local building department for footing requirements in your area.

Footings are generally as deep as the foundation wall is thick, and twice as wide. In other words, if your foundation wall is made of 6-inch blocks, the footing should be 6 inches thick and 12 inches wide.

When you dig the footing trenches, bear in mind that the top of the footing must be an even number of blocks down from the top of the foundation wall. Thus the bottom of the trench will be some multiple of 8 inches (the height of a full-sized concrete block), plus the thickness of the concrete footing, below the batter-board string lines. Sometimes you can save some extra digging by allowing for a course of 4-inch half blocks.

Lay out the footing trenches so that they will be centered under the block wall. If you will be digging the trenches by hand, use a pick and a square-sided shovel to keep the trench walls straight and the bottom as flat and level as possible. Check footing depths for level by measuring down from the string lines.

Concrete footings are normally poured directly into the foundation trench. The trick is to get the top of the footing level. The easiest way to do this is to drive a series of 2×2 stakes in the bottom of the trench, spaced 4 to 6 feet apart, with the tops of the stakes all level with the top of the footing. Later, when you pour the concrete, you'll fill the trench just to the tops of the stakes.

Reinforcing Steel

Most building codes require horizontal steel reinforcing bars in footings. You may also have to provide vertical bars up into the concrete-block foundation wall. Check local codes for foundation reinforcing requirements in your area.

Install horizontal bars in the same way as required for slab foundations (see page 56).

Vertical bars require careful planning, because they must be placed so they will be centered in the cells of the concrete blocks. Start with the corner bars. They should be set 4 inches in from the outside of the foundation wall. Position the remaining uprights every 24 inches.

Vertical reinforcing rods normally have a hook bent into the bottom end to tie them into the footing. Make the hooks so the ends will be at least 3 inches from the sides of the footing trench, and tie them to the horizontal bars. The best way to hold the upper ends of the uprights in place is to wire them to a 2×4 suspended above the footing on stakes. Brace the 2×4 every 4 feet or so to keep it from shifting when you pour the concrete. Be sure to pull the stakes just before the concrete hardens completely; the 2×4 can be left in place until the next day.

Concrete-Block Foundation Wall

When the footings are complete, you are ready to put up the foundation wall. This wall is designed to raise the building far enough above the ground so it will not be affected by soil moisture or wood-boring insects. The top of the wall should be at least 18 inches above the ground.

Blocks are nominally 6, 8, or 10 inches wide, 8 inches high, and 16 inches long. The actual measurements are about $\frac{3}{8}$ inch less, which allows for the mortar at each joint. Mortar is concrete without the gravel. To mix mortar for the blocks, use 1 part masonry cement and 3 parts washed plaster sand. You can also buy ready-made mortar mix, which comes in 60-pound sacks. Because it dries out fairly quickly, don't mix more than you can use in an hour.

The first step is to lay out the blocks on top of the footings. Locate the corners with a plumb bob hung at the intersection of the batter-board strings, then snap chalk lines between the marks. Place the blocks on the footings, spacing each one with a scrap of $\frac{3}{8}$-inch plywood, to make sure everything fits. Then remove the blocks so you can start spreading mortar.

Building a Concrete-Block Wall

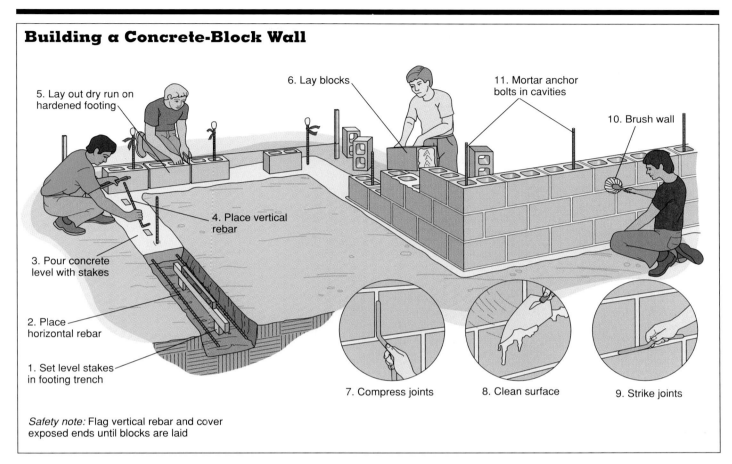

5. Lay out dry run on hardened footing

6. Lay blocks

11. Mortar anchor bolts in cavities

10. Brush wall

4. Place vertical rebar

3. Pour concrete level with stakes

2. Place horizontal rebar

1. Set level stakes in footing trench

7. Compress joints

8. Clean surface

9. Strike joints

Safety note: Flag vertical rebar and cover exposed ends until blocks are laid

Spread a layer of mortar (known to the cognoscenti as mud) under one of the corner blocks. Note that the web—the center divider of the block—is narrow on one side and thicker on the other. Always lay blocks with the thick side up.

Place the block gently in the mud so there is just a ⅜-inch layer between it and the footing. Use the end of a trowel handle to tap it square and level. Lay the remaining corner blocks in the same manner. Now you're ready to lay the blocks in between.

Stretch a piece of mason's twine along the top outside edges of two corner blocks. This will be the guide to keep the wall straight and level. Lay the blocks between cor-

ners to the string line. When you get to the last block in the course, butter both ends with mortar and slide the block carefully into place. If the mortar falls off, pull it out and try again. Continue around the footing in this manner.

Lay succeeding courses as you did the first, starting with the corners and filling in between. If you need to install horizontal steel in the foundation wall, use bond-beam blocks, which have cutouts for reinforcing bars, for one of the upper courses.

Most codes call for the hollow spaces in load-bearing walls to be filled with concrete. Even if the blocks don't have to be completely grouted, you will still have to fill

cells every 4 feet to place the anchor bolts that hold the sills to the foundation wall.

Installing Sills

Sills, also called mudsills, rest on top of the foundation wall and provide a nailing surface for the floor joists. They are usually made of 2×6 redwood or pressure-treated lumber. Lumber for sills should be carefully selected for straightness. If the building is too long for a single sill (20 feet is the longest you can buy), plan accordingly when you place anchor bolts.

To install the mudsills, first snap chalk lines on top of the foundation walls. Measure in from the outside of the wall by the same width as the sill

and snap the lines there—this will give you a reference line for keeping the sills straight. Place the sill against the anchor bolts and square across it at the bolt centerlines; then measure back from the chalk line at each bolt to locate the centers of the bolt holes. Drill holes ¹⁄₁₆ inch larger than the bolt diameter.

Slip the drilled sills over the anchor bolts and secure them with nuts and washers. Measure diagonals again to be sure the sills are square. If you've set the batter boards properly and built the foundation to the lines, the sills should be right on the money. If you're off a little, you can make small adjustments when you install the floor joists.

Floor joists, wall studs, and rafters form the skeleton of a building. If you plan ahead and avoid mistakes, framing a building can be a lot of fun—the work progresses quickly and the results are impressive. The following pages explain how to put together a frame that is strong, durable, and attractive.

Floor Framing

A typical floor frame consists of floor joists covered with a layer of plywood or oriented-strand board (OSB). This is the platform on which the rest of the structure is built. With a concrete-block wall foundation, the joists rest on the mudsills on each side and sometimes on a girder down the middle. With a post-and-pier foundation, the joists rest directly on the girders, and they don't require sills. The minimum size for floor joists is normally 2×6, but larger joists up to 2×12 can be used for longer spans.

Installing Joists

The first step in installing floor joists is to mark out their locations on the sills or girders. The first joist should be flush with the end of the building. Remaining joists are set 16 or 24 inches on center. In practice, the layout marks aren't placed in the center of the joists; move the layout left or right ¾ inch so it will be located on the edge of the joist instead. Make an X next to your mark where the joist will go so you don't forget which way you shifted the layout. If the floor joists overlap on a girder in the middle of the floor, the layout on one side of the building will have to be moved over 1½ inches to compensate.

If there will be any interior bearing walls that run parallel to the joists, locate these joists first and lay out the sills for a double joist under them. If there will be plumbing or electrical wires in the wall, space these joists 3½ inches apart so pipes and wires can be routed through the floor without cutting into the joists.

Joists should always be installed with the crown, or convex edge, up. The weight of the floor will tend to straighten the crowns. Determine the crown by sighting down the edge of each joist before placing it. Any large knots should also be on the upper side.

Floor Joists

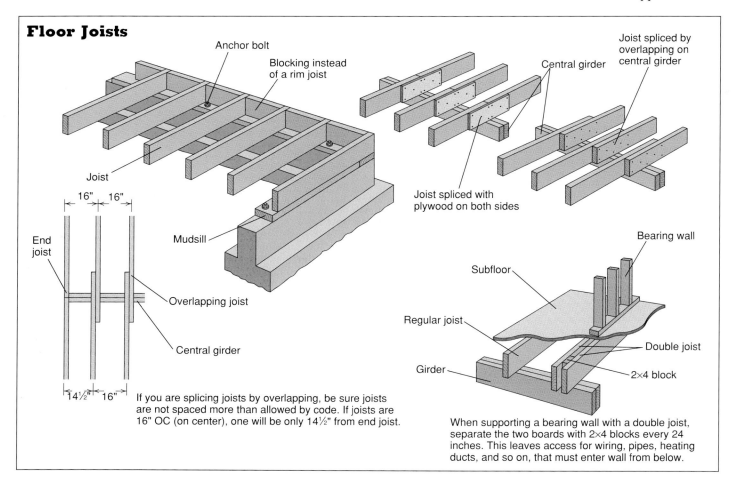

If you are splicing joists by overlapping, be sure joists are not spaced more than allowed by code. If joists are 16" OC (on center), one will be only 14½" from end joist.

When supporting a bearing wall with a double joist, separate the two boards with 2×4 blocks every 24 inches. This leaves access for wiring, pipes, heating ducts, and so on, that must enter wall from below.

Blocking

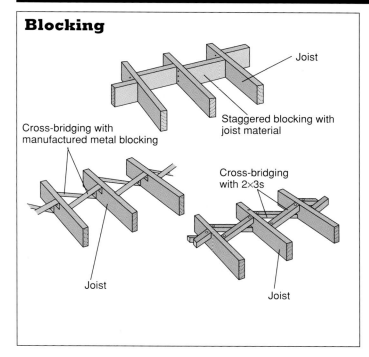

Cross-bridging with manufactured metal blocking

Joist

Staggered blocking with joist material

Cross-bridging with 2×3s

Joist

Joist

Plywood Subfloor

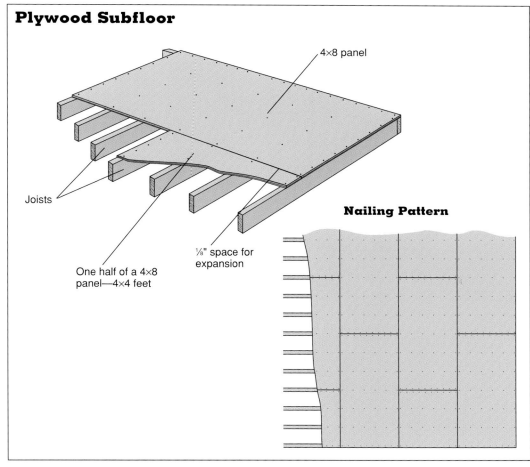

4×8 panel

Joists

⅛" space for expansion

One half of a 4×8 panel—4×4 feet

Nailing Pattern

There are two basic methods for installing floor joists. In one method, the joists stop 1½ inches from the outside of the building, and the ends are capped with a board that is the same stock as the joists, called a rim joist. This is the fastest and easiest way to build a floor. The alternative is to run the joists all the way to the outside and fill the spaces between them with blocking. This method locks the ends of the joists in place better than a rim joist, but it takes a little longer to do.

If you are using a rim joist, snap a chalk line 1½ inches in from the outside of the sills, and use the line as a guide to keep the rim joist straight as you toenail it down. Then install the joists, fastening them to the rim joist with three 16d nails and by toenailing two 16d nails into the sills or girders.

If you are using blocking between the joists, install the joists at each end of the building first, and pull a string line between their ends. Then install the remaining joists to the line. Toenail a block to the sill on each end, then nail in the joist with two nails into the sills and three nails into the ends of the blocking. Continue on down the floor, installing blocks and joists as you go. This is much easier than trying to wedge the blocking into place after the joists have been installed.

Joists can be overlapped across the center girder and nailed together (use at least a 12-inch overlap), or they can be butted end to end and spliced with 2-foot lengths of plywood or 1-by lumber. Install blocking between joists down the girder, staggering each block for easier nailing. The blocks stiffen the floor and prevent any twisting as the joists dry.

Once the joists are in place, fasten any plumbing lines to the floor joists before you cover the floor with plywood.

Installing Subflooring

At one time, floor joists were sheathed with 1-by boards, but nowadays carpenters use ⅝- or ¾-inch plywood or OSB. These materials produce a much stronger floor and a smoother surface for floor finishes.

Typical Stud Wall

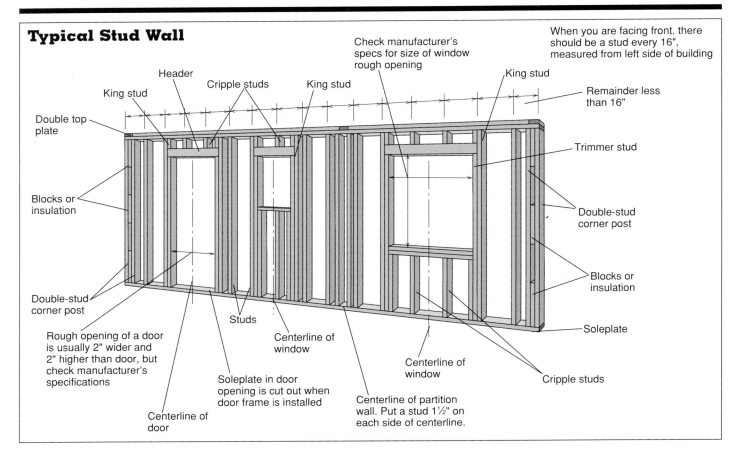

Double top plate

King stud

Header

Cripple studs

King stud

Check manufacturer's specs for size of window rough opening

King stud

When you are facing front, there should be a stud every 16", measured from left side of building

Remainder less than 16"

Trimmer stud

Blocks or insulation

Double-stud corner post

Double-stud corner post

Blocks or insulation

Soleplate

Studs

Rough opening of a door is usually 2" wider and 2" higher than door, but check manufacturer's specifications

Soleplate in door opening is cut out when door frame is installed

Centerline of window

Centerline of window

Centerline of partition wall. Put a stud 1½" on each side of centerline.

Cripple studs

Centerline of door

Install floor sheathing with the long dimension perpendicular to the joists. Snap a chalk line across the joists 4 feet in from the outside of the building to align the first course; this is more accurate than aligning the sheathing with the joists. Space plywood sheets ¹⁄₁₆ inch apart at the ends and ⅛ inch at the edges to allow for expansion. If you're using OSB, double the spacing. It's always a good idea to use construction adhesive under the floor sheathing when you install it. Glued floors are stronger, stiffer, and more fire-resistant than floors assembled with nails alone, and they are much less likely to develop squeaks later. Construction adhesives come in large tubes and are applied with a caulking gun. One 28-ounce tube will glue about three sheets of plywood. Apply only enough adhesive for one sheet at a time, or you will wind up with glue all over your shoes, your clothes, and the floor.

Fasten the panels with 8d cement-coated sinkers every 6 inches on the edges and 10 inches in the field. Angle the nails where panels butt together so they get a good bite into the joists. Snap chalk lines to locate fasteners—this is especially important when you have to jog in the middle because of lapped joists.

When the floor is all nailed off, crawl underneath and look for shiners, nails that have missed the joists. Drive them back up, pull them, and renail.

Wall Framing

Before you can begin wall framing, you need to determine the sizes of the door and window openings. Rough openings for doors are usually 2 inches wider and 2 inches taller than the nominal size of the door. Some manufacturers have different requirements, though, so check with a door supplier. Window openings are not standardized. Every manufacturer's catalog has tables showing the rough openings needed for all of its products; get a copy from a window salesperson.

Wall Layout

Layout of a stud wall proceeds in two stages. First, the wall locations are marked on the floor with chalk lines. Then the top and bottom plates are cut for each wall section and the locations of doors, windows, and studs are marked on them.

Mark for exterior walls by snapping chalk lines 3½ inches back from the edge of the floor (make the marks 5½ inches back if you are using 2×6 studs). Use two chalk lines for interior partitions to represent each side of the wall.

Cut the top and bottom plates for the walls from the longest, straightest stock that you have in your lumber pile. If you are building on a slab, use pressure-treated lumber for the bottom plate. Drill the bottom plate for the anchor bolts (see page 61).

If a wall is too long for the plate to be one continuous

piece, make the splice over the center of a stud some multiple of 16 inches from a corner of the building. Splices should be at least 4 feet away from intersecting walls. Run the plates for the long walls all the way through to the end of the building; butt the plates for the short wall up to them.

Now tack the plates together so you can lay both of them out at once. At exterior walls, tack the bottom plate to the floor and tack the top plate to it, thin edge up, so it hangs off the edge of the subfloor. To lay out plates for interior partitions, place them side by side and tack them together with a toenailed 8d cement-coated sinker at each end.

Lay out door and window openings first. On most plans, windows are positioned on the centerline of the opening. Measure back half of the rough opening width on each side of the centerline; this will be the inside face of the trimmer studs. Lay out both sides of the trimmer with a framing square, using the 1½-inch-wide tongue. Then lay out a king stud next to each trimmer. Mark king studs with an X and trimmers with a T so you don't get confused.

Next lay out the studs at corner posts and interior partitions. Mark for three studs at corner posts and two studs spaced 3½ inches apart at partitions. Make sure you lay out the flat stud on the right side of the wall.

Now lay out the rest of the studs. Just as you did for the floor joists, move the layout lines ¾ inch to the sides of the 16-inch center marks to locate the edges of the studs. Mark

each stud with an X. At window openings, mark them with a C so you'll remember that cripple studs go there. Add a cripple stud next to the trimmers at window openings to support the rough sill.

Precutting the Parts

It is usually easier to precut as many of the wall parts as you can before you start building walls. Make a cut list, showing the names, lengths, and quantities of each part needed. You can count up parts right off your plate layout. Keep your list handy so you won't get the pieces mixed up later.

Standard studs for an 8-foot wall are 92¼ inches long; most lumberyards sell studs precut to this dimension. When you add for the bottom plate and two top plates, you'll find that the finished wall will actually be 8 feet, ¾ inch tall. This allows for the thickness of the wallboard at the ceiling, plus a little clearance at the bottom of the wall. If the walls will be a different height, cut the studs accordingly, remembering to allow for the thickness of the plates.

The rest of the parts are cut as follows.

•Trimmer studs, whether for doors or windows, are typically 80½ inches long, unless you will be using 4×12 headers throughout (see box at right), in which case the trimmers are 81 inches long.

•Headers for doors are 5 inches longer than the nominal width of the door. This allows 2 inches for the jamb and shim space and 3 inches for bearing on the trimmer studs.

•Headers for windows are 3 inches longer than the manufacturer's specified rough opening width.

•Windowsills are the same length as the rough opening width. Make double sills for windows more than 4 feet wide.

•Cripple studs under windows are the trimmer length minus the rough opening height and sill thickness. For example, cripple lengths for a window with a rough opening height of 48½ inches is calculated as follows: *80½ (trimmer length) – 48½ (opening height) – 1½ (sill thickness) = 30½ inches (cripple length)*.

•Cripple studs over door openings and above and below window openings should be measured and cut after the rest of the wall is assembled, because headers can vary in width, making calculations unreliable.

•Blocks for corner posts are between 16 and 24 inches long. The exact length isn't critical. You can use scrap for corner-post blocking.

You should build up the corner posts for a wall before you assemble it. These posts are designed to provide extra support at wall intersections, as well as provide a nailing surface for the interior wall finishes. Posts for outside corners are made by sandwiching three 2×4 blocks between two studs. Fasten the blocks with at least two nails from each side. Check the ends of the post for square as you nail them together.

Posts for interior partitions are similar, but the blocking is oriented like the partition studs, so there will be a stud for

nailing on each side of the partition when the walls are in place. When the walls are erected, the end stud of the partition is nailed to the blocking.

Assembling the Walls

Wood stud walls are assembled flat on the floor, then raised into position. Two people working together can lift 16 or 20 feet of 8-foot wall. Longer or taller walls should be framed in sections and connected after they are erected. Have 16d nails on hand for most of the wall framing and 8d nails for when you have to toenail small pieces for cripple studs.

The first step is to clear the floor so you will have room to work. Remove the wall plates

Headers

Headers span the tops of openings over doors and windows to carry the weight of the roof above. They can be made of beam stock the same thickness as the walls or can be built up from 2-by material with pieces of scrap plywood between to make them the same width as the studs. The minimum size is 4×4.

You can also make headers from 4×12 stock. Even though most openings don't require headers this large, 4×12s can be installed tight to the top plate, without the small cripple studs above that would be required for smaller headers. This can save time when you frame the walls.

Marking Plates for Studs

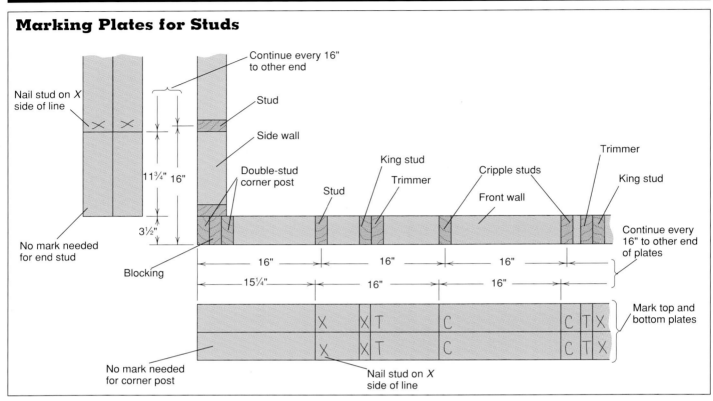

Nail stud on *X* side of line

Continue every 16" to other end

Stud

Side wall

No mark needed for end stud

11¾" 16"

3½"

Double-stud corner post

Blocking

Stud

King stud

Trimmer

Cripple studs

Front wall

Trimmer

King stud

Continue every 16" to other end of plates

16"   16"   16"

15¼"   16"   16"

Mark top and bottom plates

X   X T   C   C T X
X   X T   C   C T X

No mark needed for corner post

Nail stud on *X* side of line

that were tacked to the floor for layout. Mark each one with a letter or number and mark the floor next to it so you can replace it later. Sweep the floor clean.

Use two nails to attach the end of each stud to the plate. Nail trimmers to king studs with nails spaced 16 inches apart, staggered. Use a minimum of five nails into the ends of headers.

Start with the longest walls. Place the plates on the floor, edge up, with the layout marks facing each other, and separate them a little more than the length of the studs. Frame the window and door openings first so you won't have studs in the way when you nail the king studs to the ends of the headers. The sequence of assembly goes like this: king studs first, then

Partition and Corner Post Assemblies

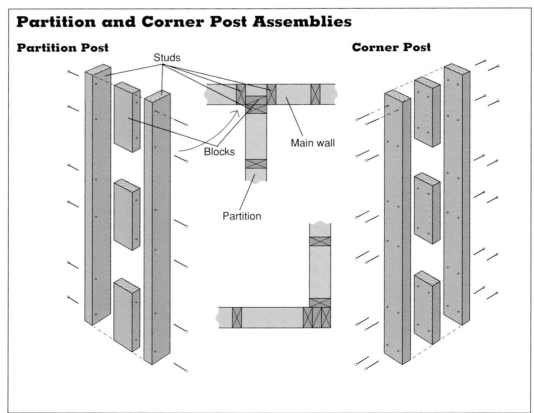

Partition Post

Corner Post

Studs

Blocks

Main wall

Partition

Header

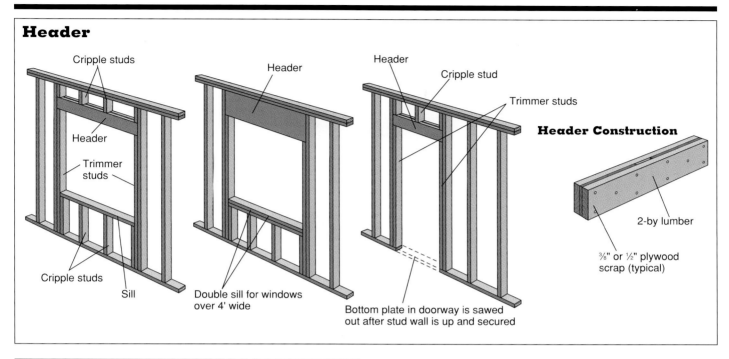

Cripple studs

Header

Header

Header

Cripple stud

Header Construction

Trimmer studs

Trimmer studs

Header

Trimmer studs

2-by lumber

Cripple studs

Sill

Double sill for windows over 4' wide

⅜" or ½" plywood scrap (typical)

Bottom plate in doorway is sawed out after stud wall is up and secured

---

Laying Out Plates

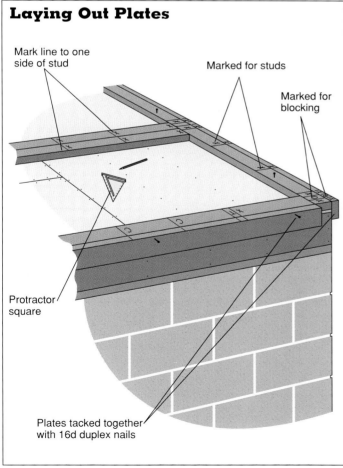

Mark line to one side of stud

Marked for studs

Marked for blocking

Protractor square

Plates tacked together with 16d duplex nails

Header Sizes

Most codes do not have tables for sizing headers over doors and windows, but only have formulas for calculating beam sizes. However, the following sizes are generally accepted rules of thumb for headers in various situations.

Location	Size of Header (4× or built-up 2×)	Maximum Span
Single Story	4×4	4 feet
or Top Story	4×6	6 feet
	4×8	8 feet
	4×10	10 feet
	4×12	12 feet
Lower Floor,	4×4	3 feet
With Floor Above	4×6	4 feet
	4×8	7 feet
	4×10	8 feet
	4×12	9 feet

Note: Increase sizes where accumulated loads concentrate on a header.

Let-In Bracing

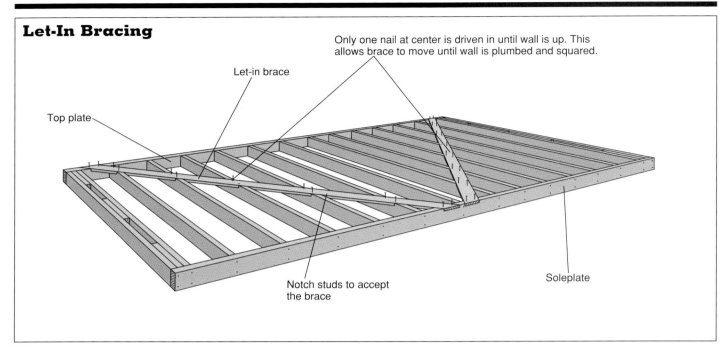

Only one nail at center is driven in until wall is up. This allows brace to move until wall is plumbed and squared.

Let-in brace

Top plate

Notch studs to accept the brace

Soleplate

trimmer studs, then headers, then cripple studs, then sills. After the openings are framed, install the corner posts and the remaining studs.

Before you install each stud, sight down its edge to see if it is crowned. Nail it into the wall with the crown up. Orienting all the studs with the crowns in the same direction will result in a flatter wall surface. Badly crowned studs should be set aside and used for bracing or cut up into blocking and cripple studs.

If the exterior walls will be covered with stucco, you should also install blocking between studs at the midpoint of the wall for at least three stud bays back from the corners and on both sides of openings. Without the blocks, the studs will be bent when the wire reinforcing for the stucco is stretched over the wall. Stagger the blocks so they will be easier to nail. Some carpenters install mid-

wall blocking regardless of the wall finishes—it helps keep the studs straight and results in a stiffer wall.

Wall Bracing

Buildings with plywood siding normally do not require bracing, but all others do. The two most widely used types are let-in bracing and metal strap bracing. Another type, called shear panel, is by far the strongest. It's used to resist earthquakes and high wind loads, or in places where conventional bracing won't fit, as on the sides of garage door openings.

Let-In Bracing

Let-in bracing is made from 1×4 or 1×6 stock and runs from the top outside corners of a wall to the bottom center to form a V shape. It is notched into the studs and plates and is prepared while the stud wall is still lying on the platform. These braces

may seem to involve a lot of detail work, but each takes just a few minutes to make, and together they add a great deal of strength to the wall.

While the wall frame is still lying on the floor, tack one of the plates to the floor by toenailing two 8d nails into the plate, stopping just short of pounding them all the way home. Push on one corner of the wall until it is square (measure diagonals to check), then tack one of the short sides to the floor. Place the brace on the wall with one end at the top corner and the other toward the middle of the wall on the bottom. A 45-degree angle is optimum, but any angle up to 60 degrees will work. The brace must cross at least three studs. Try to place the ends of the brace between studs; that way you won't be cutting through nails when you make the notches. Tack the brace to the wall with two 8d nails.

Mark the studs and plates where the brace crosses them. Then pull the two 8d nails and remove the brace. Set a circular saw blade to cut ⅛ inch deeper than the thickness of the brace and cut every line you marked on the studs and plates. Next, make a series of closely spaced cuts between these lines and knock out the waste with a hammer. Clean up the bottoms of the notches with a chisel.

Place the brace in the notches, letting the ends run wild. Pound in two 8d nails wherever the brace crosses a stud or plate. Trim the brace flush with the outer edges of the plates.

Metal Strap Bracing

This type of bracing comes in two types. One is a flat strap that is applied to the studs after the walls have been erected and plumbed. The other type is an L-strap; one leg of the L fits into a shallow saw

Metal Strap Bracing

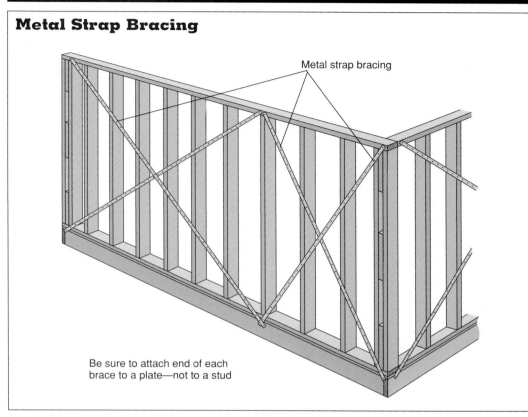

Metal strap bracing

Be sure to attach end of each brace to a plate—not to a stud

cut in the studs. Both types are as strong as let-in bracing, and they are considerably easier to install. Metal braces come in 10- and 12-foot lengths.

Metal strap braces come with holes for 8d nails drilled every 2 inches. They must always be put up in crossed pairs, as illustrated. To attach them, first make sure the walls are squared and plumb, then nail the brace to the top plate and nail it to every framing member it crosses. To install metal L-straps, set a circular saw for a ¾-inch-deep cut and saw through all the lines. Place the brace in the saw cut and nail it in place.

Shear Panel

In some areas, particularly in earthquake or hurricane country, you may be required to install shear panel at strategic

Metal L-Strap Bracing

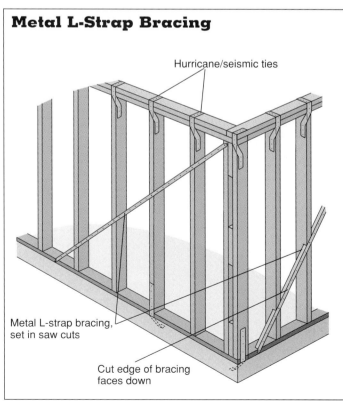

Hurricane/seismic ties

Metal L-strap bracing, set in saw cuts

Cut edge of bracing faces down

Shear Panel

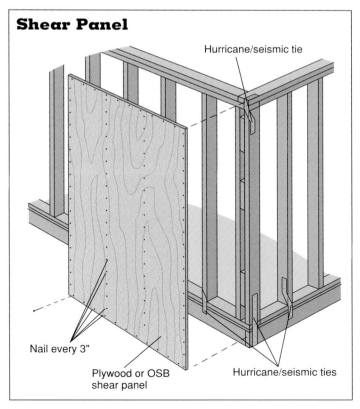

Hurricane/seismic tie

Nail every 3"

Plywood or OSB shear panel

Hurricane/seismic ties

Bracing a Stud Wall

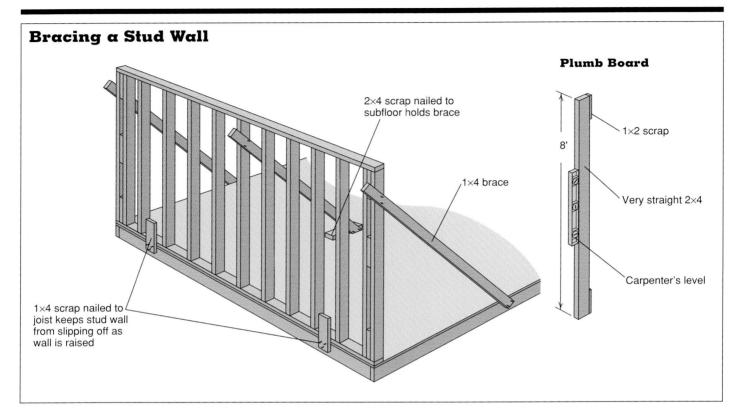

2×4 scrap nailed to subfloor holds brace

1×4 brace

1×4 scrap nailed to joist keeps stud wall from slipping off as wall is raised

Plumb Board

1×2 scrap

8'

Very straight 2×4

Carpenter's level

locations on your building. Shear panel consists of sheets of plywood or OSB nailed to the studs from soleplate to top plate. It makes a very strong brace. Plywood siding, when properly fastened, serves the same purpose.

Sheets of shear panel are normally applied vertically, with the long dimension parallel to the studs. You can also put them up horizontally, but this requires a row of blocking in the middle of the wall for nailing along the horizontal edges of the sheets.

It is important to use the right number and type of fasteners when you install shear panel. The normal requirement is an 8d every 6 inches around the edges of the sheets and every 12 inches on intermediate studs, but in highly stressed walls the spacing may

be reduced to as little as 3 inches between nails.

Sometimes metal anchors are cast into the foundation to work in conjunction with the shear panel to resist uplift. These can be metal straps or heavy anchor bolts connected to special hold-down brackets bolted to the studs.

Raising the Walls

Use an assistant when you raise the walls into position. The two of you can lift the wall together, then one person can hold it steady while the other nails up temporary bracing.

Before you lift a wall, nail some pieces of scrap 2×4 to the rim joists, with the ends sticking up above the floor, to keep the wall from sliding off when you lift it.

Raise the long outside walls first. Start by sliding the wall

into position, until the bottom plate lines up with the chalk line on the floor. Use hammer claws to lift the top plate up enough to slip a block of scrap wood under it—this will give you room to get your fingers under the wall when you lift. Keep your back straight and lift with your legs when you pick up the wall.

Once the wall is standing, install temporary 2×4 bracing to hold it there. It's a good idea to lean the wall out an inch or two when you brace it—that way you will have some clearance to raise the end walls. Later you can loosen the braces to tie the corners together.

Place the first brace from the top end of the wall to the outside of the end joist. If you are building on a slab, run the brace to the outside of the slab and secure the bottom end to a stake in the ground.

Once the wall is braced, tap the bottom plate into position with a hammer, lining it up exactly with the chalk lines on the floor, and nail it down with 16d nails every 16 inches. Be sure the nails penetrate into the joists below, not just the floor sheathing.

Raise the rest of the walls, tacking the corners together as you go and installing temporary bracing. Don't worry about getting them exactly plumb yet; you'll do that after all the walls are in place and the cap plates installed.

Installing Cap Plates

Some carpenters prefer to install cap plates while the walls are still flat on the floor, but you'll get a better fit if you do it after the walls are up. Select the straightest possible lumber for cap plates. The cap plates on the end walls overlap the side walls, tying the corners together at the top. Cap plates on any interior walls must tie onto an exterior wall. Any joints in the cap plate must be at least 4 feet from joints in the top plate.

Nail the cap plates down with 16d nails staggered every 16 inches, plus two nails at the ends of each piece. Try to keep the nails over the studs so you won't hit them if you have to drill through the plates later for wires or pipes.

Once the cap plates are on and all the walls are tied together, you can plumb and straighten the building.

Plumbing and Straightening Walls

Carpenters call this part of the job plumb and line. All the corners are plumbed and braced, then all the walls are straightened to a string line. This is the final phase preparatory to framing the roof. Before you start, you should make up a plumb board as shown in the illustration on page 70. Make it the same length as the wall height.

Plumb the walls, starting with the longest one first. If the wall doesn't have permanent braces, nail a temporary 2×4 brace to the inside of the wall; it should remain there until the roof is framed and the exterior siding is applied. Continue around the building, plumbing each wall and bracing it as you go. Try to get within ¼ inch of plumb at every corner.

Long walls have a tendency to bow in the middle, so they

have to be straightened before you can frame the roof. To do this, nail small blocks of 1-by scrap along the side of the top plates, a few inches from each end. Don't drive the nails all the way home. Stretch a string line from nail to nail along the top plate; the blocks will serve to space the string out to bridge any irregularities. Then check the space between the string line and the top plates at several points along the wall with another piece of scrap 1-by. If the spacing varies more than ⅛ inch, bend the top plate into line with a push stick and install a brace from the top of a stud to a block nailed to the floor. Check to see if there are any voids between the bottom plate and the floor. Wedge in builder's shim, held by construction adhesive, under any studs where gaps occur.

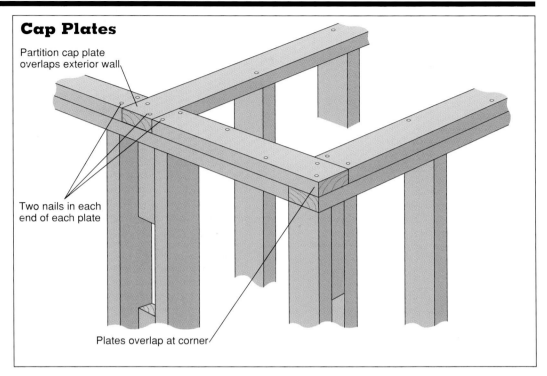

Cap Plates

Partition cap plate overlaps exterior wall

Two nails in each end of each plate

Plates overlap at corner

This window opening has two headers because two separate windows will fit into the space. Note that the bottom plates are pressure-treated lumber because they rest directly on the concrete floor.

ASSEMBLING THE ROOF

Roof assembly proceeds as follows: the ceiling joists are cut and nailed in place, then the rafters are installed. The gable ends are framed in next, and outriggers are installed to support the overhangs. The fascia board is nailed to the rafter tails and outriggers. Then, as the final step, the whole roof is sheathed with plywood or 1-by boards.

You'll be doing most of the work from ladders or standing on top of the walls, so exercise caution to prevent accidents. Use sensible ladder practice, and don't leave tools lying on the walls where they could fall and injure someone.

Ceiling Joists

Besides supporting the finished ceiling, ceiling joists tie the walls together to resist the outward thrust of the rafters. Since the ceiling joists will be nailed to the rafters, you should lay out the positions of rafters and joists at the same time. Rafters and ceiling joists are normally spaced 24 inches on center, but in areas with heavy snow loads you may have to reduce the spacing to 16 inches on center.

Mark the rafter locations just as you did for the floor joists and studs, starting from the same corner. Don't forget to shift the layout ¾ inch from the centerlines so you can see the marks when you install the rafters. Mark one side of the line with an *R* (for rafter) and the other side with a *J* (for joist) so you can keep track of which piece goes where.

Cut the ceiling joists the same length as the building's width. Toenail them to the top plates; their ends should be flush with the outside of the walls. The upper corners of the joists may extend above the rafters—these will be trimmed off later, after all the rafters are in place. Don't install the last two joists at the ends of the building yet. The gable studs must be installed first so the joists can be nailed to them.

Ceiling joists are normally made from 2×6s, which will not span the full width of most houses. Splices in ceiling joists must be supported in the middle by interior walls or beams. When joists are butted together over an interior wall, the ends must be tied together

Roof Framing

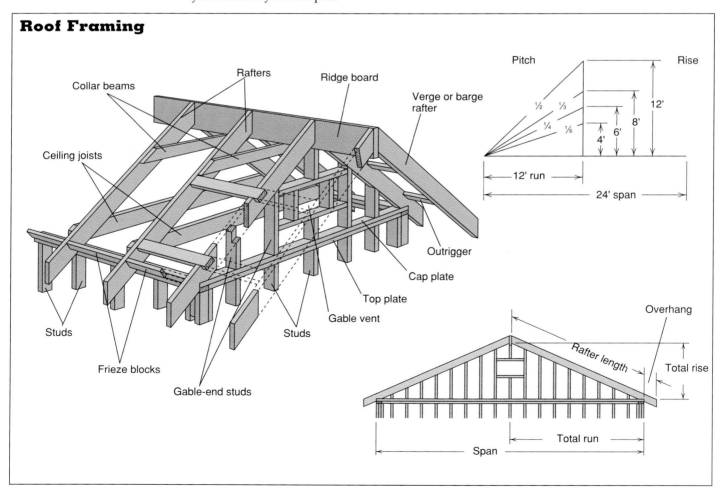

Cutting Rafters

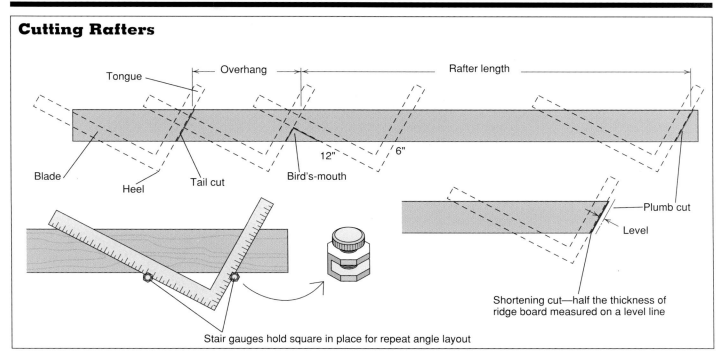

Stair gauges hold square in place for repeat angle layout

with 2-foot strips on ⅜-inch plywood on both sides.

If you want a vaulted ceiling, you will have to tie the rafters together somehow. The usual way to do this is with collar beams, which are nailed across each pair of rafters (see illustration at left). Collar beams must be placed no higher than one third of the way down from the ridge. Nail them to the rafters with three 16d nails in each end.

Roof Framing

Roof framing is arguably the most demanding part of the carpenter's job. It takes practice and skill to frame a complicated roof. A complete discussion of roof framing, with hip, valley, and jack rafters, is beyond the scope of this book. Simple roofs, however, are not difficult to build. Here is how to construct the basic gable roof, the most common residential type.

Roof Pitches

A gable roof consists of common rafters, a ridge board, gable-end studs, and ceiling joists or collar beams. The angle, or pitch, of the roof is based on the principle of the right triangle. The horizontal distance from the outside of the building to a point directly below the ridge forms one side of the triangle; it is called the run. This distance is typically half the width of the building. The vertical distance from the top of the rafter at the wall line to the top of the ridge is the second side of the triangle; it's called the rise. The rafter is the third side, or hypotenuse, of the triangle.

The slope of a roof is defined by the number of inches of rise per foot of run. Thus, a rafter that rises 6 inches for every 12 inches of run has a 6 in 12 pitch; a rafter that rises 4 inches every foot has a 4 in 12 pitch.

Another method of describing roof pitches is the ratio of the rise to the width of the building, but it is rarely used nowadays. For example, a roof whose rise is one fourth of the span, or building width, would be called a ¼-pitch roof.

If the roof illustrated at left has a span of 8 feet and a total rise of 4 feet, it has a 6 in 12 pitch. The principles for laying out and cutting the rafters apply to all roofs, regardless of the width of the building or the slope of the roof.

Cutting Rafters

All the angled cuts for the rafters are laid out with the framing square. A handy aid for laying out rafters is the stair gauge fixture, available in hardware stores. It is a slotted button with a thumbscrew that clamps it to the square. A pair of gauges can be affixed to the framing square so they touch the edge of the rafter;

with the buttons in place, you can slide the square up and down the rafter to lay out the cuts without having to reset the angle each time.

Plumb Cut

The first step is to lay out the plumb cut, which forms the end of the rafter that fits against the ridge. First sight down the rafter to see which way it is crowned, and place it on sawhorses with the crown facing away from you. The crowned edge will be the top of the rafter.

For the 6 in 12 pitch illustrated, place the framing square on the right-hand end of the rafter, with the heel facing you and the tongue on the right. Align the 6-inch mark on the outside of the tongue and the 12-inch mark on the outside of the blade with the edge of the board. Draw a line along the side of the tongue. This line represents the center of the ridge. (You'll trim the rafter later to allow for the thickness of the ridge board.)

If you had chosen a different pitch, you would have used a different mark on the tongue, but you would have kept the 12-inch mark on the blade. The 12-inch mark remains constant; pitch adjustments are made by moving the tongue.

Bird's-Mouth

This cut, sometimes called a seat cut, is an angled notch near the bottom of the rafter. It's made so the rafter can sit firmly on the wall.

To locate the position of the bird's-mouth, you first have to determine the length of the rafter from the tables on the framing square. There are six lines on the rafter tables, but you need only be concerned with the first one. It says "length of common rafters per foot run," followed by a series of numbers. Look along the scale on the outside of the blade until you find the rise of your rafter pitch (in this case, the 6-inch mark). Directly below the 6, on the first line of the tables, are two numbers—in this example, 13 42. This means that for every 12 inches of run, the rafter will be 13.42 inches long. To find the length of the rafter, multiply the number in the table by the run (in feet). For the playhouse roof shown, the run is half the building width, or 4 feet. The rafter length, therefore, is *13.42 × 4 = 53.68 inches*. Round this number off to the nearest ⅛ inch, or 53⅝ inches.

If the run is not an even number of feet, you will have to convert the dimension to decimal fractions to calculate

rafter length. Measure the run in inches, divide by 12, and multiply by the number on the rafter tables. A pocket calculator is helpful here.

Measure along the rafter length from the tip of the plumb cut and mark the edge of the rafter. Holding the 6- and 12-inch marks, line up one edge of the tongue with the mark and draw a line there. This mark represents the outside of the building. Now, still holding the 6- and 12-inch marks, slide the square to your right, toward the plumb cut. Position it so the distance from the outer edge of the board to the line you have just drawn is 3½ inches. Draw another line along the edge of the blade. This line represents the top of the wall. Cut out the section formed by these lines, using a circular saw to cut into the corner of the notch and finishing the cut with a handsaw.

Tail Cut

To make the tail cut, slide the square down toward the rafter end for as much overhang as your plans call for, and mark along the outside of the tongue. If you are too close to the end of the board to line up the 6- and 12-inch marks, turn the square end for end and mark from the other edge of the rafter. The angle should match that of the plumb cut. Eventually you will nail a fascia board or rain gutter to the exposed rafter ends.

Shortening Cut

This step is simple, but must not be overlooked. Rafter lengths are first measured as if they butted together at the

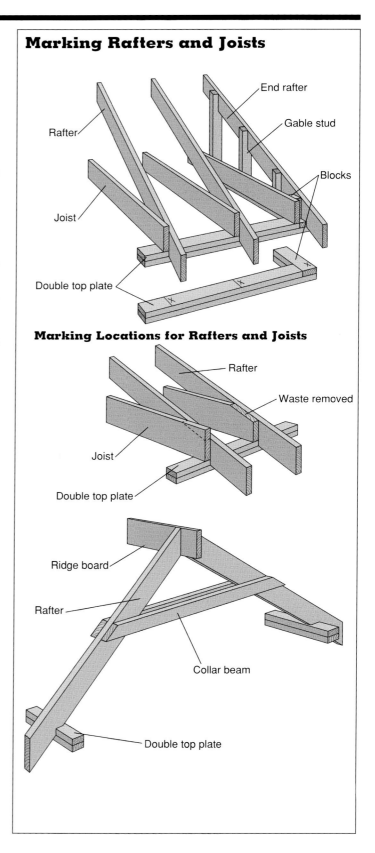

Marking Rafters and Joists

Rafter · End rafter · Gable stud · Blocks · Joist · Double top plate

Marking Locations for Rafters and Joists

Rafter · Waste removed · Joist · Double top plate

Ridge board · Rafter · Collar beam · Double top plate

Roof Framing Details

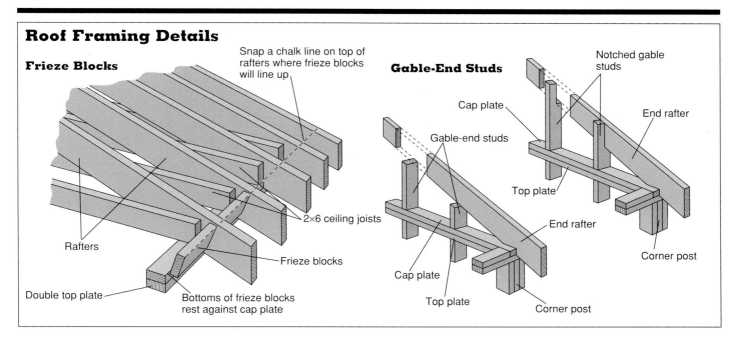

Frieze Blocks

Snap a chalk line on top of rafters where frieze blocks will line up

Rafters

2×6 ceiling joists

Frieze blocks

Double top plate

Bottoms of frieze blocks rest against cap plate

Gable-End Studs

Notched gable studs

Cap plate

Gable-end studs

End rafter

Top plate

End rafter

Corner post

Cap plate

Top plate

Corner post

---

peak of the roof. Most roofs, however, are constructed with a ridge board to make it easier to put up the rafters.

For the rafters to fit properly, half the thickness of the ridge board must be trimmed from each rafter at the plumb cut. If you are using a 1-by ridge, which is ¾ inch thick, trim ⅜ inch from the rafter end. If you're using 2-by stock, trim off ¾ inch. The cut must be parallel to your original cut; lay it out with the framing square.

Once the first rafter has been accurately cut, use it as a pattern to lay out and cut the remaining rafters.

Ridge Board

Now cut the ridge board to length. The ridge is made from stock that is one size wider than the rafters (for example, use a 2×8 ridge for 2×6 rafters). It should be as long as the building, plus the overhangs on each end. Lay out the rafter locations on it to match the layout on the walls.

If you have to make the ridge in two pieces, put the splice where two rafters meet, not in the spaces between rafters.

Barge Rafters

In addition to the rafters that rest on the walls, you will also need four barge rafters. These rafters fit at the outside of the gable overhang and are supported by outriggers. They are similar to the common rafters, but don't have the bird's-mouth cut.

Installing Rafters

Putting up a roof frame can be tricky until the first few rafters are installed. You have to hold up five pieces at once (four rafters and the ridge board) and nail them together at the same time. You'll need at least one helper.

Start by propping the ridge up to the proper height. To do this, hold two opposing

rafters in place above one end wall, with a scrap of the ridge board between them, and measure down from the bottom of the ridge to the top plate. Make temporary supports from scrap 2×4 and nail them to the end walls to hold the ridge at that elevation. Then install the rafters over the end walls, with three 8d toenails into the plates and two 16d nails at the ridge.

Once the first four rafters are up, the rest are easy. Pull a string line along the top of the ridge so you can keep it straight as you work. Install each rafter by nailing the bottom end to the plates first, then nailing the top end to the ridge. Fasten each rafter to the adjoining ceiling joist with three 16d nails.

When the rafters are all up, install the frieze blocking. These blocks fit between the rafters and are made from the same stock. Install them just beyond the ends of the ceiling joist so the top edges are flush with the tops of the rafters

and the bottom edges rest against the cap plate. Snap a chalk line on the top of the rafters to keep all the blocks in a straight line.

Gable-End Studs

Once all the rafters and frieze blocks are in place, install the studs in the triangular gable ends. These studs should line up with the wall studs below. Each must be custom-cut. Make the studs by first cutting a piece of 2×4 slightly longer than needed. Plumb it and mark it along the underside of the rafter. Then either angle-cut it or notch it by making a 1½-inch-deep angle-cut on the edge of the stud, followed by a rip cut from the end of the stud to the first cut. Toe-nail the stud to the cap plate and the rafter with 8d nails.

If you will be installing gable-end vents, frame openings in the center of the gables to fit. Then install the ceiling joists left out earlier, nailing them to the gable studs.

Roof Sheathing

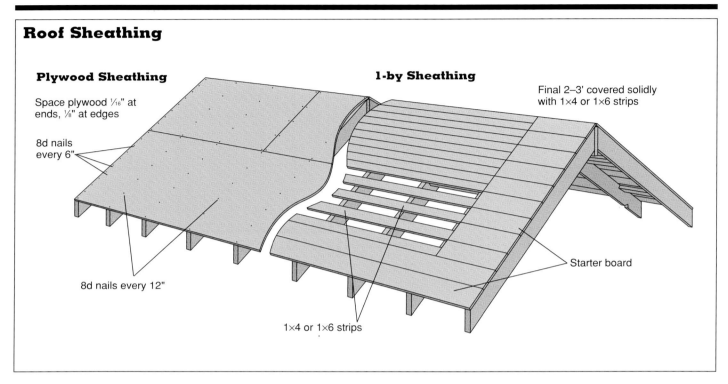

Plywood Sheathing

Space plywood 1/16" at ends, 1/8" at edges

8d nails every 6"

8d nails every 12"

1-by Sheathing

Final 2–3' covered solidly with 1×4 or 1×6 strips

Starter board

1×4 or 1×6 strips

Outriggers

Outriggers provide the support to extend the overhang at the gable ends. They consist of lengths of 2×4 attached flat and extending from the last rafter before the wall end out to the barge rafter. The rafter above the wall must be notched where the outrigger crosses it. Outriggers should be spaced no more than 4 feet apart. If you will be using plywood roof sheathing, try to center the outriggers at the plywood joints. When all the outriggers are in place, nail the barge rafters to the protruding ends and the ridge board, using 16d nails.

Fascia

A fascia board protects the ends of the rafter tails and provides support for the edge of the roof sheathing. It can be 1-by or 2-by material, but should be at least one size

wider than the rafter stock so it will cover the ends of the rafters completely. For the neatest appearance, use kiln-dried stock of a wood species that will hold up to the weather. Install the fascia with hot-dipped galvanized (HDG) nails to avoid corrosion.

Put up the horizontal fascia first. You will get better nailing for the roof sheathing if you bevel the top edge to match the roof pitch. Make any splices at the end of a rafter tail, with the ends of the pieces beveled at 45 degrees so they overlap at the joint. The gable-end fascia boards are applied over the barge rafters. Measure and cut them to length, using a framing square to lay out the angle-cuts, just as you did for the rafters. For a neater appearance at the corners, miter the ends of the fascia boards at 45 degrees where they meet.

Vents

A vent at each end of the house, close up under the eaves, allows air to move freely through the attic and prevents the buildup of condensation. There are several types available, usually made of metal or wood. Louvered vents will keep out the weather; they should be covered with screening on the back side to keep birds and insects out of the attic.

Roof Sheathing

There are several ways to cover a roof frame, each dictated by eave style and roofing materials. Which one you choose will depend on your tastes, your budget, and the type of roofing material you plan to use.

•Starter board, which is usually 1×6 or 1×8 boards with shiplap or tongue-and-groove edges, is frequently used to

cover the overhangs of a house to give the exposed underside of the eaves a finished appearance. Starter board can be applied over the entire roof if the sheathing will be exposed on the inside of the building.

To install starter board, first snap a chalk line on the tops of the rafters so you can get the first piece straight. Position the chalk line a board width up from the rafter end. Nail the first piece to the chalk line, then install the rest until the overhang is covered. The starter board at the gable ends should be cut so it is flush with the outside of the fascia and covers half the rafters over the end walls. This leaves nailing space for the sheathing that will cover the rest of the roof.

•Spaced sheathing, which is either 1×4 or 1×6 boards, is used under wood shingle and shake roofs. The boards

Laying the Roof

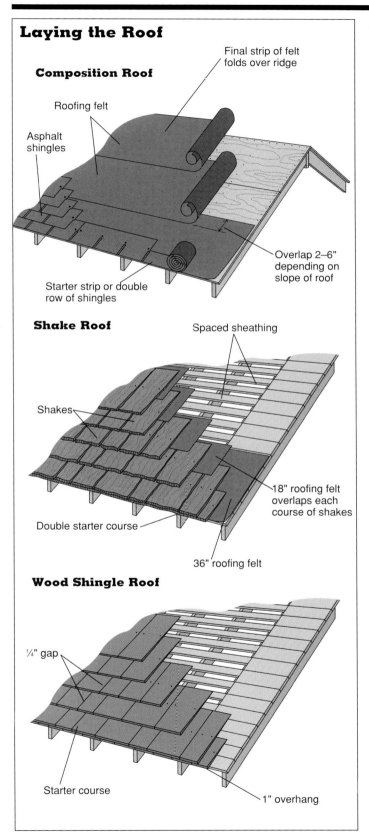

Composition Roof

Final strip of felt folds over ridge

Roofing felt

Asphalt shingles

Overlap 2–6" depending on slope of roof

Starter strip or double row of shingles

Shake Roof

Spaced sheathing

Shakes

18" roofing felt overlaps each course of shakes

Double starter course

36" roofing felt

Wood Shingle Roof

¼" gap

Starter course

1" overhang

are installed approximately one board width apart—the spaces allow air to circulate around the shingles to reduce cupping and splitting. Up near the ridge, for the last 2 feet or so, the boards are butted tightly together to provide a solid nailing surface for the last few courses of shingles or shakes.

• Plywood and oriented-strand board (OSB) are the sheathing materials of choice for other types of roofing. When you install plywood or OSB sheathing, start from the bottom of the roof and work up. Snap a chalk line 4 feet up the roof to get the first row straight. Apply the sheets with the long dimension perpendicular to the rafters. Space plywood sheets ¹⁄₁₆ inch apart at the ends and ⅛ inch at the edges; double the spacing for OSB. Fasten them down with 8d nails at 6 inches on center on the edges and 12 inches in the field. If you use OSB sheathing, remember to put the textured side up so you won't slip.

Roofing

The easiest types of roofing to install are wood shingles, shakes, and asphalt composition shingles. Other types of roofing require specialized tools and techniques and are better left to professionals. There is no magic to installing shingles, as long as you follow the manufacturer's recommended procedures and use basic safety sense.

The minimum slope for shingled roofs is normally 4 in 12, although slopes as low as 3 in 12 can be shingled successfully with additional underlayment. The maximum slope

you can safely navigate without safety equipment is 6 in 12. If your roof is steeper than this, call in a roofing contractor.

Wood Shingles

Wood shingles are normally installed over spaced sheathing. An underlayment is not required unless you live in an area where ice dams form around the eaves in winter; then you should install a waterproof membrane under the shingles. It should extend from the eaves up the roof to a point 2 to 3 feet inside the exterior walls.

Shingles are always installed from the eaves up, with each course of shingles overlapping the ones below. The amount of each shingle that is left exposed to the weather, known as the exposure, depends on the roof slope and the size of the shingles.

The first course of shingles should be two layers thick and should extend about 1 inch beyond the roof sheathing. Each shingle should be fastened with two 4d galvanized nails placed about ¾ inch in from the sides and 1 inch above the butts of the next row. You may have to use shorter nails over the eaves to keep the points from poking through the sheathing at exposed overhangs. Leave a ¼-inch gap between shingles for expansion. Offset joints at least 1½ inches between courses. When you get to the ridge, let the shingle ends stick up in the air and trim them all off at once with a circular saw. Then cover the ridge with factory-made hip-and-ridge shingles, working in from both ends toward the middle.

Shakes

Shakes are installed much like wood shingles, except an underlayment of roofing felt is required. The first layer is a 36-inch-wide strip of 30-pound felt along the eaves. The next course is an 18-inch-wide strip of 30-pound felt laid so its bottom edge is twice the shingle exposure up from the edge of the roof. Succeeding courses of 18-inch-wide felt are put down so the exposed width is the same as the shake exposure. When you install the shakes, tuck the top end of each one under the felt above. If you have installed the strips of felt carefully, you can use the preprinted lines on the surface to keep the rows of shakes straight. Nail shakes with galvanized 6d nails placed so the next course will cover the nail heads by about 2 inches.

Composition Shingles

Composition-shingle roofs must be installed over a solid roof deck. Before you load the shingles onto the roof, put down a layer of 15-pound roofing felt, lapping horizontal edges at least 3 inches and vertical seams 4 inches. Most manufacturers print excellent installation instructions on the shingle wrappers—if you follow them carefully, you shouldn't have to worry about the roof for another 15 or 20 years.

Flashings

Nearly all roof leaks can be traced to problems with the flashings. Flashings are required around chimneys, skylights, vent pipes, roof valleys, and anywhere a roof meets a vertical wall surface. No matter what type of roofing material you use, it is absolutely essential to pay special attention to flashing details.

Flashings can be made of copper, lead, aluminum, stainless steel, or regular steel coated with zinc, terneplate, or special paints. Galvanized steel and aluminum flashings are by far the most common because of their low cost and relatively long life. The other types will last even longer, but they're significantly more expensive. Many of the most common shapes can be purchased directly from building-supply outlets, but if you need a custom shape, a sheet-metal shop can make it for you.

Valley Flashing

You must put flashing along the entire length of a valley before roofing over it. Valley flashing for composition-shingle roofs is usually a double layer of roll roofing. Wood shingles and shakes require a metal W-flashing, which has a rib in the center to keep water flowing down one side of the roof from getting under the shingles on the other side of the valley.

Vent Flashing

Vent flashing is required wherever a vent pipe, flue, or other conduit penetrates the roof. Buy vent collars that are designed to fit the pipe and suit your particular roofing

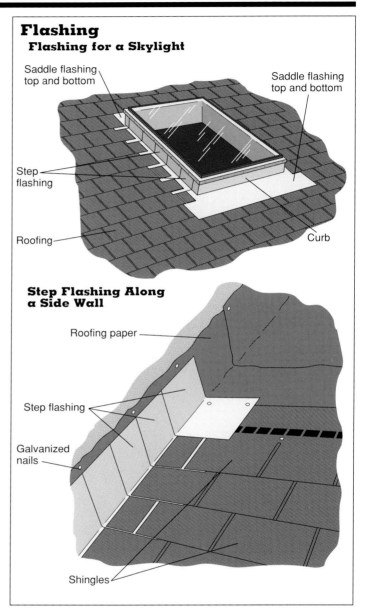

Flashing
Flashing for a Skylight

Saddle flashing top and bottom

Saddle flashing top and bottom

Step flashing

Roofing

Curb

Step Flashing Along a Side Wall

Roofing paper

Step flashing

Galvanized nails

Shingles

material. Install the flashing as if it were a shingle, with the top tucked under the shingles above and the bottom lapped over the shingles below.

Step Flashing

Use step flashing wherever a sloping roof intersects a vertical surface. Install a piece of step flashing with every course of shingles. The bottom edge should overlap the shingle below so that any water that makes its way under the shingles will be shed back onto the roof surface. The vertical edge of the flashing can be covered with the wall siding, or you can use a counter flashing with a spacer strip behind it. This is the preferable option, because when it comes time to reroof, the step flashings can be replaced without removing the siding.

Installing Chimney Flashing

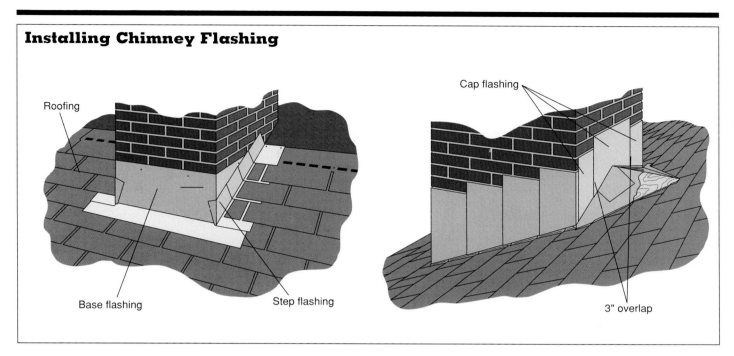

Roofing

Base flashing

Step flashing

Cap flashing

3" overlap

Flashing in Valleys

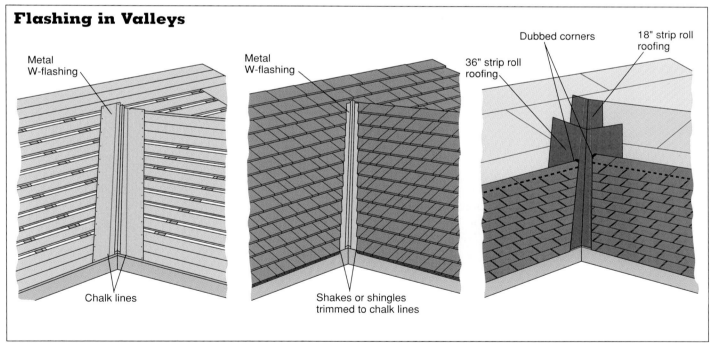

Metal W-flashing

Chalk lines

Metal W-flashing

Shakes or shingles trimmed to chalk lines

Dubbed corners

18" strip roll roofing

36" strip roll roofing

Saddle Flashing

Saddle flashings wrap around the top and bottom sides of skylights and chimneys and extend partway around the corners. It's best to have these made up by a sheet-metal shop. Install the lower saddle over the shingles. Use step flashing up the sides, then install the top saddle and shingle over it. At chimneys, the flashing is covered by a cap flashing embedded in the mortar joints of the masonry.

Roof-to-Wall Flashing

If the joint between the roof and a wall is horizontal, the flashing is installed as a continuous piece that is bent at an angle to match the roof slope. The bottom edge should cover the shingles by 5 or 6 inches. The top edge, like step flashing, is covered by siding or counter flashing. This type of flashing can be installed before the roofing if the bottom half is bent upward to slip the roofing under it, but it is easier to install it after the roofing.

FINISHING AND TRIM CARPENTRY

Building stairs, installing windows and doors, adding interior trim—no doubt about it, these important finishing touches require finesse. There's no room for hammer dents at this stage. But if you've paid close attention to details thus far—keeping things plumb, level, and square—you'll likely find that finish work goes fairly easily.

Stairs and Railings

In the most advanced forms, building stairs and handrails is the most demanding aspect of the carpenter's art. Stairs don't have to be fancy to function well, though. A homeowner with basic carpentry skills and a pocket calculator should be able to build a simple set of straight stairs.

Remember that building codes are very strict about stair construction. There are specific regulations governing maximum slope, landing requirements, and handrail design. The instructions that follow will conform to most codes, but you should still check with the building department for the rules that apply in your area.

Stair Layout

The first step when designing a stair is to determine the number of risers required. The maximum allowable riser height is 8 inches, but a riser between 6 and 7½ inches will result in a safer, more comfortable stair. To determine the number of risers, measure the distance between top and bottom landings, divide by 7, and discard any fractions. Then divide the total rise by this number to find the exact height of each riser. Thus, if the distance between landings is 108 inches, the calculation goes like this: *number of risers required = 108 ÷ 7 = 15.43, or 15 risers; riser height = 108 ÷ 15 = 7.2, or 7³⁄₁₆ inches.*

Next, determine the width of the treads. There is a specific relationship between riser height and tread width. In general, one riser plus one tread should equal 17 to 18 inches. Steep stairs should be at the lower end of this range, so a stair with an 8-inch riser would have a 9-inch tread, for a total of 17 inches. A stair with a gradual slope is at the other end of the spectrum: a 6-inch riser, for example, would require a 12-inch tread.

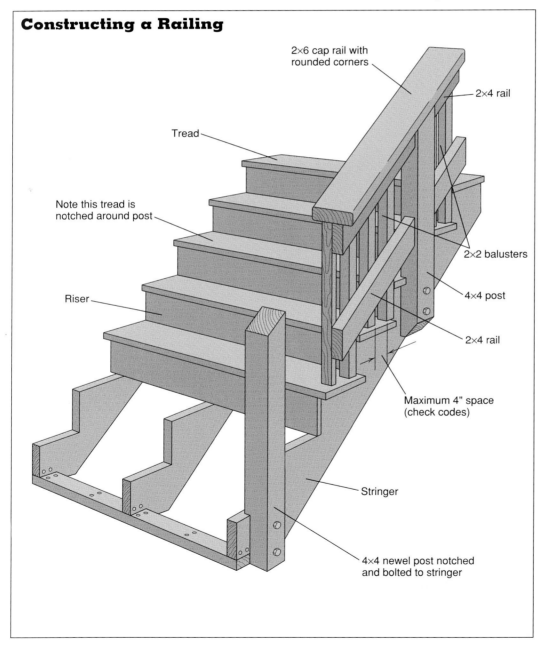

Constructing a Railing

2×6 cap rail with rounded corners

2×4 rail

Tread

Note this tread is notched around post

2×2 balusters

4×4 post

2×4 rail

Riser

Maximum 4" space (check codes)

Stringer

4×4 newel post notched and bolted to stringer

Installing Stringers

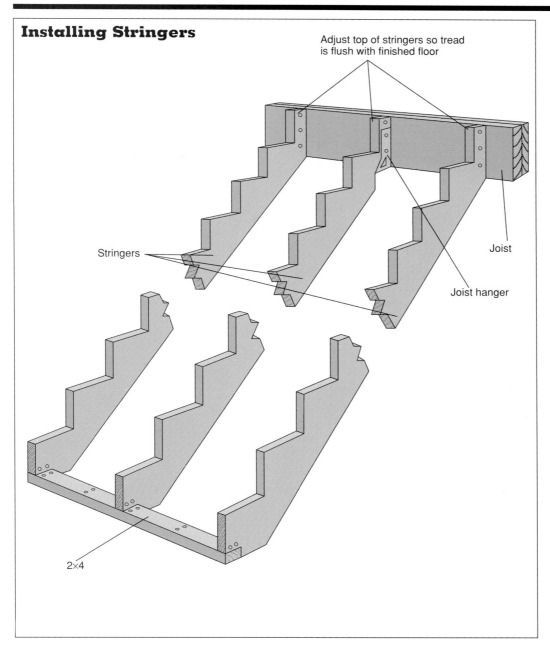

Adjust top of stringers so tread is flush with finished floor

Stringers

Joist

Joist hanger

2×4

Marking a Stringer

Notch for 2×4

Waste

Step off 14 risers

Framing square

10½"

Cut thickness of tread from bottom of stringer

Stringer

7¼"

Other stairs fall somewhere in between. In our example, a riser of 7³⁄₁₆ inches would have a tread about 10½ inches wide.

Notched Stringers

Stair stringers, which are notched for the treads and risers, should be made from 2×12 stock of the best quality you can find. Two stringers are usually adequate for stairs up to 32 inches wide; use 3 or 4 stringers for wider stairs. To lay out the notches, set a framing square with the riser height on the tongue and the tread width on the blade lined up with the edge of the stringer stock. Scribe the first notch onto the stringer, then move the square over to mark the rest of the notches. Pay special attention to the cuts at the top and bottom of the stair (it may help to make a full-scale drawing on a sheet of plywood to help you visualize how the parts go together). To make the bottom step the same height as the others, trim the bottom of the stringer by the thickness of the tread material, and cut a notch for an anchor cleat made from a 2×4. At the top, lay out the upper end of the stringer so it can be supported by metal hangers or a ledger cleat.

Cut out the stringer with a circular saw, using a handsaw to get into the corners of the notches. Hold the stringer up in place to check the fit, and make adjustments as necessary. Use the first stringer as a pattern to cut the remaining stringers.

Treads and Risers

Exterior stairs normally have 2-by lumber treads and open risers. Interior stairs usually have enclosed risers. Interior stairs that will be carpeted can have treads and risers made of ¾-inch plywood or oriented-strand board (OSB).

The finished treads are normally cut about 1 inch wider than nominal tread run; this leaves a small overhang at the riser for more stable footing when ascending the stairs. Riser boards, when they are used, are cut to the same riser dimension you used to lay out the stringers.

If your stair uses riser boards, install them first. Fasten them to the stringers with construction adhesive and 8d nails, keeping the nails away from the points of the notches so they don't split. Glue and nail treads to the stringers, working from the bottom up, and nail through the risers into the backs of the treads.

Railings

The design possibilities for railings are endless, but all stair railings must conform to basic safety requirements. There must be a smooth gripping surface no more than 2 inches thick located between 34 and 38 inches above the stair nosings, the front edges of the treads. Most codes require that railings be constructed so that a 4-inch sphere cannot be passed through at any point (some codes allow for 6-inch spaces). The railing must be capable of withstanding a horizontal force of 20 pounds per lineal foot.

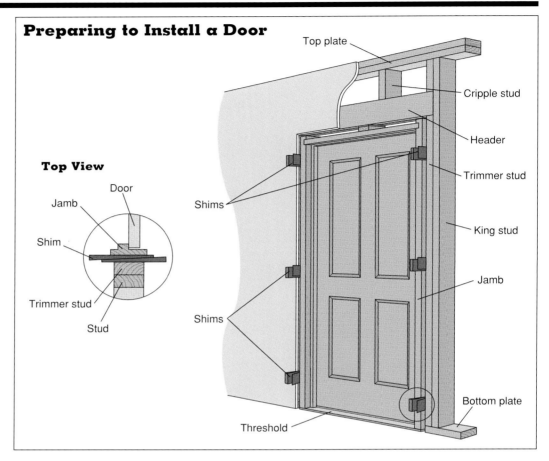

Preparing to Install a Door

Top View

Door
Jamb
Shim
Trimmer stud
Stud

Top plate
Cripple stud
Header
Trimmer stud
King stud
Jamb
Shims
Shims
Bottom plate
Threshold

One simple railing that meets all these criteria is shown in the illustration on page 80. It consists of 4×4 posts for structural support, a 2×6 cap rail that overhangs the posts to provide a handgrip, and a balustrade of 2×2 pickets supported by 2×4 rails.

To build the railing, first install the posts. Space them 3 to 6 feet apart. Cut a ¾-inch-deep notch on the bottom ends to fit over the stair stringers, and fasten them to the stringers with two hex-head lag screws in each post. Install the 2×6 cap rail next, rounding the corners to prevent splinters. Then toenail the intermediate rails to the posts and fasten the pickets to them. Try to space the pickets equally.

Guardrails that protect the edges of decks and balconies can be constructed much like stair railings, except that the minimum height for guardrails is usually 36 inches.

Doors

The easiest way to put in a new door is to buy a prehung unit, which comes complete with frame, hinges, and pre-bored holes for the lock. Exterior prehung doors also have the threshold, door-bottom seal, and weather stripping installed at the door shop.

When you order a prehung door, you will have to specify the door size, the direction of the swing, the lock hardware preparation required, and the width of the jamb. This last dimension should be equal to the overall thickness of the finished wall, plus ⅟₁₆ to ⅛ inch extra to bridge irregularities in the wall surface.

Installing Prehung Doors

When you framed the rough openings for the doors, they were approximately ½ inch larger than the door frame. The extra space allows some room to make minor adjustments to get the doors plumb. The gap between frame and trimmer studs is filled with tapered wood shims placed under the fasteners. You can buy door-hanging shims at a lumberyard or use roofing shingles.

The first step is to check the rough opening for plumb. You can be sure it will not be perfect. Check both sides of the opening, holding a level against the trimmer studs and also against the face of the wall. Note which way the opening is off—when you install the door, you'll make the appropriate allowances.

Now nail shims to the trimmer studs on the hinge side of the opening, placing them where the top and bottom hinges will go. The shims are normally put up in pairs, with the tapers facing in opposite directions, but if the trimmer stud is twisted, you may have to use one or more shims facing the same direction to keep the jamb square to the wall. Slide the shims back and forth, checking the level until they are exactly plumb to each other.

Remove the door from the jamb (pull the hinge pins), and slide the jamb into the opening. Before you fasten it to the shims, check the top of the frame for level. If the hinge side is low, slip a shim under the jamb leg to raise it up. Now hold a level against the edge of the jamb and adjust the frame until it is plumb. It may not be perfectly flush with the wall; if it isn't, try to split the difference top and bottom. When the hinge jamb is plumb in both directions, fasten it to the trimmer studs with two 8d finishing nails through each pair of shims.

Now put the door back in the frame. Use more shims to secure the latch side of the frame, placing them at the top and bottom and behind the latch strike. Check the gap

Fitting Door Casing

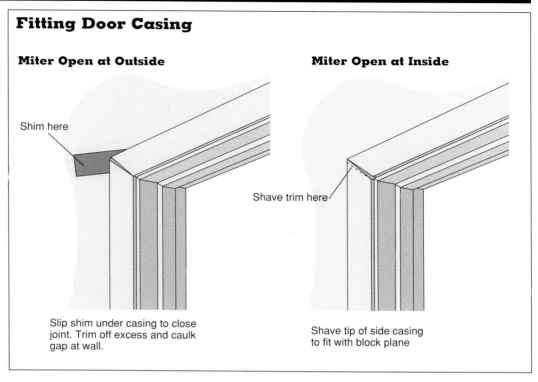

Miter Open at Outside

Shim here

Slip shim under casing to close joint. Trim off excess and caulk gap at wall.

Miter Open at Inside

Shave trim here

Shave tip of side casing to fit with block plane

between door and frame frequently as you work, trying to maintain a uniform reveal all around. You should also check that the door seats against the door stop from top to bottom. Again, you may have to fudge the jamb in or out a bit. Don't worry if it's not exactly flush with the wall; the door casings will hide the discrepancy.

Finally, install additional shims as necessary. You should place them at three points on each side of the door, with additional shims as required to straighten bowed jambs or support heavy doors. When you are finished, saw off the protruding ends of the shims.

Door Trim

The trim around the door, called door casing, covers the gap between door jamb and trimmer studs. It also serves a structural purpose, tying the jamb to the trimmers. Door casing is normally applied so its inner edge is about ³⁄₁₆ inch back from the face of the door jamb. This setback is called the reveal. Door casing is fastened with pairs of nails every 16 inches, with 3d finishing nails into the jamb and 6d nails into the framing. Keep the nails 2 or 3 inches back from the ends of the pieces to reduce splitting.

To install the casing, cut the top piece first, making 45-degree miters at the corners. The short side of the casing will be ⅜ inch longer than the finished opening to allow for the ³⁄₁₆-inch reveal on each end. Fasten it in place as described above.

Then cut the side casings, mitering the top ends. Hold the first piece in place and check the fit at the miter. If the jamb is not on the same plane as the wall, the joint probably won't fit very well. For a miter

that is open at the inside corner, use a block plane to shave the end of the side casing until it fits. Miters that are open at the outside corner can usually be fixed by tapping a shim behind the miter, which moves the casings away from the wall slightly. Cut off the excess shim and caulk the resulting gap before you paint the walls and trim. To keep the miters tight, drive small brads through the corners of the casing.

Windows

Residential windows are made of aluminum, vinyl, or wood. The aluminum and vinyl types come with a flange around the outside, called a nailing fin, that is used to fasten the window to the framing. Wood windows are installed similar to doors, with shims and finishing nails.

Preparing the Opening

Before installing windows, cover the outside of the rough opening with flashing paper to keep water from working its way in around the window. Flashing paper is usually a sandwich of kraft paper, asphalt, and reinforcing fibers, and it comes in 6- or 12-inch-wide rolls.

First, staple a strip of flashing across the windowsill, with the ends extending at least 6 inches past the opening on each side. Staple only the top edge of this piece—later you will tuck a moisture barrier under the bottom edge. Then run a piece up each side of the window, with the bottom ends overlapping the first strip. Another strip goes over the top of the opening after the window is installed.

Installing Aluminum and Vinyl Windows

Aluminum and vinyl windows are the easiest to install. Begin by placing the window in the opening and holding it there from the outside. Have an assistant block the window up off the rough sill with shims and adjust it in the opening until the gap between framing and window is uniform on all sides. Nail the top flange with a 1½-inch roofing nail at each end. Check the window for square by measuring diagonals. If it's out of square, your assistant can gently pry one of the lower corners over until the diagonals are equal. Then drive nails partway in at the bottom corners of the window to hold it,

and check to see that it opens and closes smoothly. If it does, install the rest of the nails, spacing them about 12 inches apart all around the nailing fin. Then staple up the last strip of flashing paper so it covers the flange at the top of the window.

Installing Wood Windows

Two types of wood windows are currently available. One is the traditional wood window, which has a wood frame and a sloping sill that protrudes past the side jambs on the outside. The other type is clad on the outside with aluminum or vinyl for weather protection. This latter type usually has a nailing fin similar to the one on all-metal or vinyl windows. It lacks the wide sill of traditional windows.

To install a traditional window, you will first have to cut the ears to length. These are the portions of the sill that extend past the side jambs on both sides of the window. The purpose of the ears is to provide a place to terminate the bottom ends of the trim covering the side jambs; cut the ears so they will extend beyond the side trim about ½ inch.

Place the window into the rough opening and wedge some shims in at the sides to hold it in place temporarily. Cut tapered blocks to match the sill slope, and slip them under the bottom of the window until the sill is level. Fasten the sill down with nails long enough to penetrate into the framing below. Shim the side jambs plumb, check the window for square, and secure it in place with nails

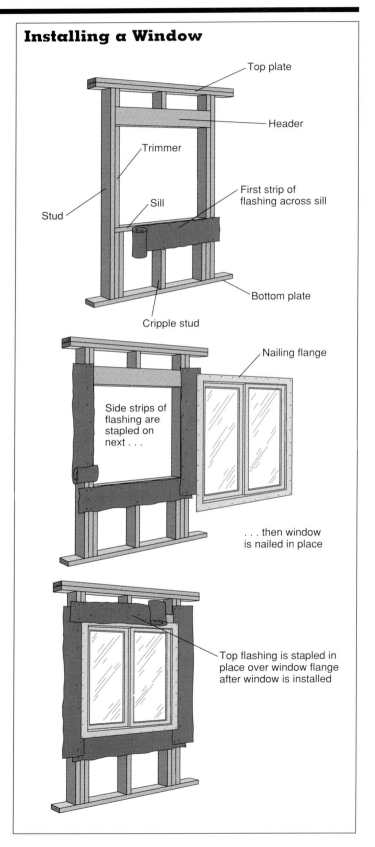

Top plate

Header

Trimmer

First strip of flashing across sill

Stud

Sill

Bottom plate

Cripple stud

Nailing flange

Side strips of flashing are stapled on next . . .

. . . then window is nailed in place

Top flashing is stapled in place over window flange after window is installed through the shims. Saw off the protruding ends of shims and sill blocks when you are done.

Clad windows are installed in much the same way as aluminum and vinyl windows. Staple flashing paper around the opening, square up the window, and nail the flange to the framing. Then, working from the inside, shim between the window jamb and the framing. Nail the jamb to the rough sill, trimmer studs, and header.

Siding

There are many ways to cover the outside walls of a house. This section explains how to install three of the most common types of wood siding: plywood, board-and-batten, and horizontal boards.

Wrapping the House

Before you install any kind of siding, the building should first be wrapped with some kind of moisture barrier. Moisture barriers prevent any water that makes its way through the siding from getting into the wall. Moisture barriers are impermeable to liquid water, but they contain microscopic pores that allow water vapor to escape to the outside.

The most common type of moisture barrier is an asphalt-saturated paper that comes in 3-foot-wide rolls. It is usually called building paper or black paper. Fifteen-pound roofing felt, which also comes in 3-foot rolls, works just as well, although it is harder to handle. Both are installed the same way.

Apply moisture barriers with staples or roofing nails. Staples are faster and easier to use, but if you are putting up the paper on a breezy day, nails will secure it better.

Always start from the bottom of the wall and work your way up. The lower edge of the first course should extend approximately 1 inch below the mudsill at the bottom of the wall. Succeeding courses should overlap the paper below by at least 3 inches. Vertical splices should overlap a minimum of 6 inches.

At window openings, tuck the paper up under the bottom strip of flashing you installed with the window. At the sides and tops of windows, the paper overlaps the flashing and is trimmed tight to the window frame.

Sometimes it's tough to put up building paper without getting wrinkles in it. The trick is to fasten only the top two corners of each piece when you stretch it out on the wall. Then pull the paper snug and fasten the ends securely. Staple the paper to the intermediate studs last.

Housewraps

Another type of moisture barrier, known as housewrap, has been developed recently. It reduces air infiltration into houses. Air-infiltration barriers can improve the energy efficiency of a building significantly. This synthetic material comes in rolls up to 10 feet wide—enough to cover a whole wall of a single-story house in one piece.

Housewraps are tough—they're practically tearproof—so they are more resistant to

accidental punctures than building paper or felt. Because they are so wide, they go up fast. As you might expect, housewraps are more expensive than other moisture barriers.

Housewraps can be installed much like building paper, but in order to get the most from them, you need to take extra care to seal all potential air leaks. Seams, for instance, are covered with a special tape. Most manufacturers provide detailed installation instructions with their products; read and follow them if you use a housewrap to protect your building.

Plywood Siding

Plywood is the least expensive type of siding. It's also the easiest to install. It comes in 4-foot-wide sheets that are 8, 9, or 10 feet long. The long edges of the sheets are rabbeted so they overlap at the seams. Plywood siding usually has a rough-sawed texture, sometimes with a grooved pattern cut into it. Standard thicknesses are ⅜ inch and ⅝ inch.

Plywood siding comes in three grades. Utility, the lowest grade, has surface defects that have been filled with putty. It's suitable for utility buildings. Custom grade has the defects filled with football-shaped wood patches. It's fine for paint-grade work, but the patches may show through a transparent stain. Premium grade is free of surface defects.

If your house is built with standard 92¼-inch studs, you should buy siding in 9-foot sheets. The extra length lets you cover the whole wall from the underside of the raf-

ters to the bottom of the floor framing.

Start putting up the plywood panels at one corner of the building. Cut them to length, allowing an extra inch to lap onto the foundation. Plumb the first sheet carefully when you nail it up. The leading edge of the sheet should fall just past the center of a stud. Don't worry about how it fits at the corner—the corners will be covered with trim later.

Fasten the panels with hot-dipped galvanized (HDG) 8d nails every 6 inches on the edges and 12 inches in the field. If you laid out the wall studs carefully, every seam will land directly over a stud. If you find you don't have a stud where you need it, nail a piece of backing into the wall from the back side and fasten the siding to that.

Leave a ⅛-inch gap between the siding and the door and window frames (fill the gap with caulk before you install the trim). When you have to cut a sheet around a door or window, it helps to make a sketch of the sheet with all the cut dimensions marked on the drawing. Also note which side of the sheet is which so you won't cut from the wrong edge. It's surprising how easy it is to make a mistake, so take your time, and double-check your work before you cut.

At the gable ends, you will have to cut the tops of the sheets at an angle to follow the roof pitch. You will also have to notch out for the outriggers. To determine the correct angle, use a level to draw plumb lines on the wall at the edges of the sheet, then

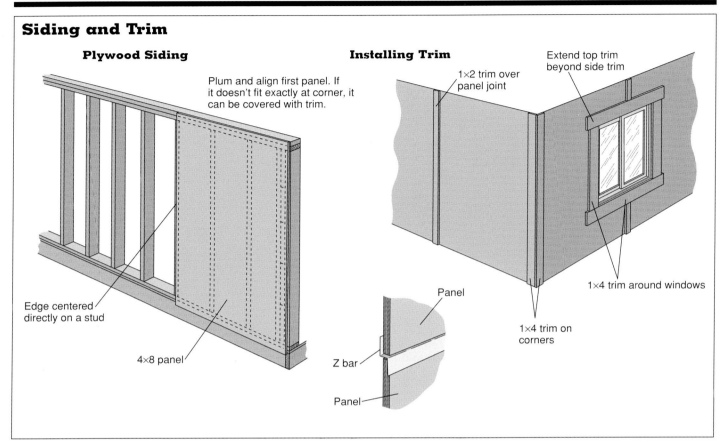

Plywood Siding

Plum and align first panel. If it doesn't fit exactly at corner, it can be covered with trim.

Edge centered directly on a stud

4×8 panel

Installing Trim

1×2 trim over panel joint

Extend top trim beyond side trim

1×4 trim around windows

1×4 trim on corners

Panel

Z bar

Panel

---

measure the height at each end. Again, make a sketch and measure carefully. Where the plywood rests on the lower wall panels, install strips of metal flashing, called Z bars, to keep water out of the joint.

Trim

Trim for plywood paneling is usually 1×3 or 1×4 material. Back-prime the trim before applying it. Square-cut trim around doors and windows is usually better than mitered trim because miters tend to open up when exposed to the weather. The trim over the openings should extend past the side pieces to keep water out. Use HDG 8d box nails to fasten the trim. As an added safeguard against the elements, apply caulk to door

and window jambs before attaching the trim. Corner trim will fit better if you nail the two pieces together before you nail them to the building.

Board-and-Batten Siding

Board-and batten siding has an attractive, rustic look. It's not difficult to install, but does require some planning and forethought for best results.

Siding boards for a board-and-batten job vary in width from 1×8 to 1×12. Battens are normally 2 to 3 inches wide. Trim around windows, doors, and corners can be the same material as the battens, but wider trim usually looks better.

Because the boards are installed vertically, you will need horizontal backing in the framing so you'll have something to nail to. You can install rows of blocking between the studs on 2-foot centers, or you can nail furring strips over the studs. At minimum, use 1×4 furring; 2×4s will hold nails better. If you use furring strips, put them up before you install the doors, windows, and moisture barrier. Install short pieces of furring around door and window openings to provide backing for the trim.

Before you install any boards, make a story pole. This is a 10- or 12-foot strip of wood with the centerlines of the joints between boards marked on it. Hold it against the wall, noting where the

joints will fall. The object is to avoid joints too close to windows, doors, and corners. Siding boards are normally installed with a ½-inch expansion gap between them, but you can adjust the width of the gap if necessary to get a better siding layout. You may want to install a narrow board over the top of an opening if it will improve the pattern of battens on the rest of the wall. When you're satisfied with the layout, mark the joint locations on the wall with a lumber crayon.

The secret of success when installing wide siding boards is to allow for expansion and contraction caused by seasonal weather changes. Boards that are fastened on both edges

Board-and-Batten Siding

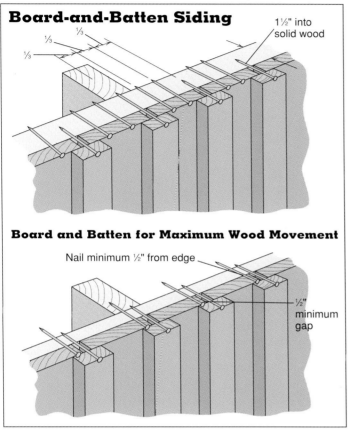

1½" into solid wood

⅓ ⅓ ⅓

Board and Batten for Maximum Wood Movement

Nail minimum ½" from edge

½" minimum gap

may split down the middle if they shrink. Instead, place the nails near the center of the boards or down one edge only. The battens, which are nailed through the gap between boards, will hold the free edges tight to the wall.

If you plan to paint or stain the siding boards, it's a good idea to preseal both sides with whatever finish you will be using. Sealing the back side will help prevent cupping later. When the sealer is dry, fasten the boards to the wall with HDG or other corrosion-resistant nails, following the layout you put up earlier. The nails should penetrate at least 1½ inches into the framing.

Once all the boards are up, install the trim around the doors, windows, and corners. Then install the battens, making sure the nails go into the gaps, not the siding boards.

Horizontal Boards

Another option for wood siding is to install boards that are mounted horizontally on the wall. Horizontal siding is more weatherproof than vertical boards because the pieces are shaped to shed water to the outside. This type of siding emphasizes the horizontal lines of the house, which may be an important design consideration.

When the siding arrives on the job site, divide the boards into three piles. The best, straightest boards go into the first pile. These will be used on the most visible walls. The second stack is for boards with minor defects; these will be used to cover the rest of the walls. Put any pieces that are split, badly bowed, or otherwise unacceptable into the third stack, and

Horizontal Siding

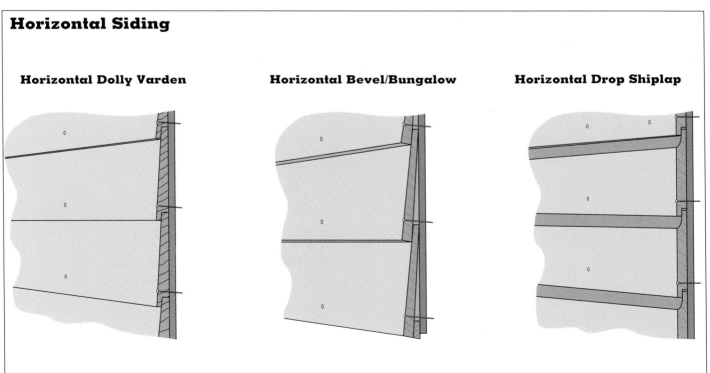

Horizontal Dolly Varden

Horizontal Bevel/Bungalow

Horizontal Drop Shiplap

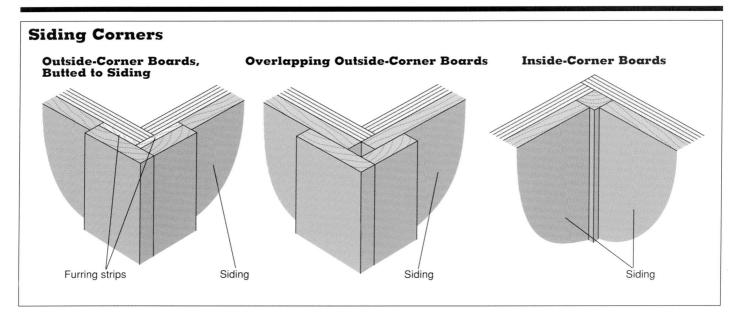

Siding Corners

Outside-Corner Boards, Butted to Siding

Overlapping Outside-Corner Boards

Inside-Corner Boards

Furring strips

Siding

Siding

Siding

ask the supplier to replace them. Prime the siding boards on both sides to reduce cupping.

There are two ways to install horizontal siding boards. In one system, the boards are nailed to the walls and the trim is applied over them. This is the fastest and easiest method, but with some styles of siding, you will be left with hundreds of small gaps where the siding boards pass under the trim. These crannies are ideal habitat for bugs and dirt. The second method eliminates this problem by installing the trim first and butting the siding to it. This makes for a neater installation, but requires more careful workmanship. All the cuts around the trim are exposed, so any sloppy fits will be there for all to see.

To install the siding, first cover the walls with building paper or housewrap. Snap chalk lines on the walls to mark the studs. If you will be installing the trim after the siding, set the first board to

a chalk line snapped on the wall. The bottom edge should overlap the foundation by 1 inch. If you use clapboard siding, nail a spacer strip under the bottom edge of the first course. Then install the rest of the boards, leaving a ⅛-inch expansion gap between them. Bowed pieces can be bent straight as you fasten them. Nail the boards on the bottom only, using HDG 8d box nails or specialized ring-shank siding nails. Position each nail just above the top edge of the board below. Install the trim over the siding in the normal way.

If the siding will be butted into the trim, you'll have to install the trim first. The trim should stand out slightly past the face of the siding—you may have to put up plywood spacer strips to accomplish this. At inside corners, nail a strip of 1×1 into the corner where the walls meet. When you fit the siding boards to the trim, measure carefully. You can make fine adjustments by

shaving the ends of the pieces with a block plane. Caulk under the ends of each piece as you install it to seal the joints. All end-to-end joints should land over studs. Try to scatter these joints in a random pattern over the wall so they won't be too obvious.

Interior Trim

Now that you have the outside of the building covered up, you can proceed with finishing the interior. Plumbing lines must be installed, electrical outlets and lights wired, heating and cooling systems put in, and walls covered with wallboard or paneling. Once this work is out of the way, you can start on the interior finish carpentry.

Baseboards

After installing the interior doors and casings (see page 82), baseboard trim is the next step. If you set yourself up properly, installing baseboard goes quickly. First, go around

all the walls and mark the stud locations on the floor. If you can't find the studs, there is a handy little gadget called a stud finder that will locate them for you electronically. Find two or three studs, then use a tape measure to mark out the rest on 16-inch centers.

Now set up a miter saw on a pair of sawhorses, with all the baseboard stacked alongside. You will cut the pieces for one room at a time. Stand in the doorway of the first room and draw a quick floor plan of the walls. This is just a schematic to tell you how to make the cuts—it doesn't have to be even close to scale. Then go around the room and measure each wall section, marking the dimensions on your drawing.

Coped Corners

Baseboards that meet at an outside corner are mitered. Most of the walls in a room, however, meet at an inside corner. If you miter the baseboard in these corners, the joint is sure to open up when

Joints in Wood

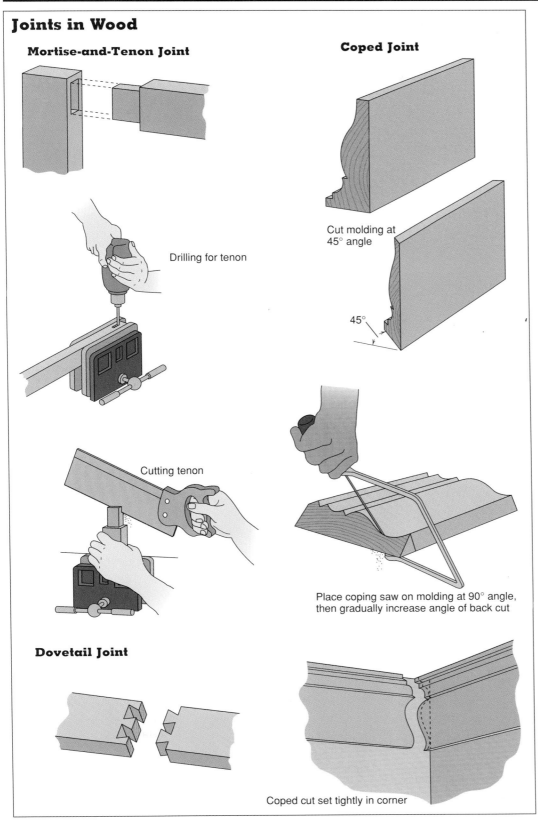

Mortise-and-Tenon Joint

Drilling for tenon

Cutting tenon

Dovetail Joint

Coped Joint

Cut molding at 45° angle

45°

Place coping saw on molding at 90° angle, then gradually increase angle of back cut

Coped cut set tightly in corner

you nail the pieces to the wall. Baseboard with a simple rectangular shape can be butted at inside corners. If the baseboard has a molded profile, it's best to use a coped joint. One end of each piece is cut square and run into the corner, and the piece that butts into it is cut with the reverse profile and fitted to it.

This sounds like a difficult process, but the procedure is really very simple. First cut the piece to the length shown on your sketch, with a 45-degree back-bevel on the coped end (most right-handed people find it easiest to cope the right-hand end of the piece). Then use a coping saw to cut the profile, following the line along which the bevel meets the face of the molding. Angle the saw back a little so you will get a tight fit at the exposed face.

When you have all the pieces for a room cut, fasten them in place with finishing nails. If you coped the right-hand ends of the baseboard, you will be working from right to left around the room. Nail the baseboard with pairs of nails at each stud mark. If the floor will be carpeted, space the baseboard off the floor about ⅜ inch so the carpet can be tucked under it (a scrap of baseboard makes a good spacer).

Sometimes you will have to install a very short piece of baseboard next to a door opening. Don't try to nail it in, or you will be sure to split it. Instead, just put some glue on the back side and press it into place. Cope the piece that fits up to it to hold it there until the glue dries.

Opposite page: Top left: Most of the framing is complete for this single-story home, which has a concrete slab foundation. The walls have been framed and braced and roof trusses have been installed over them. Top right: The walls of this home are attached to the slab foundation with conventional anchor bolts and additional anchoring devices at the corners to prevent uplift from strong winds or earthquakes. Center left: Permanent bracing for this wall consists of a 1×4 set into notches cut into the studs and plates. The brace must extend from the bottom plate to the top plate. Center right: The top plates on this gable wall are pitched because the roof trusses, called scissors trusses, are sloped on the bottom to create a sloped ceiling. Bottom left: The rafter length of this roof is too long for single boards, so each rafter consists of two boards that overlap each other over a supporting wall. The ventilation holes drilled into the rafter blocks will allow air to circulate freely above the roof insulation. Bottom right: A pattern rafter with a bird's-mouth cut into it will be used to mark and cut the second rafter.

This page: Top: Roof sheathing is being installed over the trusses. The open areas along the eaves and gable end will be sheathed with a more attractive material than the oriented-strand board used for most of the roof, because the sheathing in those areas will be visible from below. Center: Roof sheathing is completed. To provide a more interesting roof line, a small gable, partially visible here, was built over the left window after the primary roof was sheathed. Some of the wall area was sheathed at the time it was framed; it can be built out with furring studs and the final sheathing to create a deeper profile, as with the garage door opening. Bottom: The first stage of roofing is to cover the sheathing with underlayment, usually 15-pound felt paper. The roofing for this house will be tiles, which require furring strips. The windows have been installed, with flashing applied around the framed opening. The 1-by trim for the windows is installed before the stucco is applied.

REFERENCE TABLES

Nailing Schedule for Residential Construction

Codes specify the size, number, and placement of nails for framing connections. This table cites specifications typical of most codes. However, check with local officials to be sure these figures are in compliance with applicable codes.

Connection	No. and Size of Common Nail	How Nailed
Joist to sill or girder	3 8d	Toenailed
Built-up girders and beams of 3 members	20d 32" OC[1]	Facenailed
Bridging to joist, each end	2 8d	Toenailed
Ledger strip (ribbon) at each joist	3 16d	Facenailed
1×6 subfloor, to each joist	2 8d	Facenailed
Wider than 1×6 subfloor	3 8d	Facenailed
2" subfloor to joist or girder	2 16d	Facenailed
Soleplate to joist or blocking	16d @ 16" OC	Facenailed
Soleplate to stud	2 16d	End nailed
	4 8d	Toenailed
Top plate to stud	2 16d	End nailed
Doubled studs	16d @ 16" OC	Facenailed
Built-up corner studs	16d @ 24" OC	Facenailed
Doubled top plates	16d @ 16" OC	Facenailed
Top plates, laps, and intersections	2 16d	Facenailed
Continuous header, 2 pieces	16d @ 16" OC[2]	Facenailed
Continuous header to stud	4 8d	Toenailed
Ceiling joists to plate	3 8d	Toenailed
Ceiling joists, laps over partitions	3 16d	Facenailed
Ceiling joists to parallel rafters	3 16d	Facenailed
Rafter to plate	3 8d	Toenailed
1" brace to each stud and plate	2 8d	Facenailed
1×8 sheathing or less, each bearing	2 8d	Facenailed
Wider than 1×8 sheathing, each bearing	3 8d	Facenailed

OC[1] at top and bottom, staggered

OC[2] along each edge

Using Span Tables

The tables on the next page look complicated because they must account for the different strengths of various grades and species of lumber. The numbers across the top of each chart represent the range of strengths found in commercially available lumber. Some typical lumber grades and their strengths are listed at the bottom of this column to make it easier for you to use the charts. More complete tables are printed in code books and carpentry manuals.

There are separate tables for floor joists, ceiling joists, and rafters. To use a table, determine the span from your working plans, decide whether to space the members on 16- or 24-inch centers, then compare the various span limits for different sizes of lumber within the same column. Choose a vertical column that corresponds to the strength of lumber you plan to use. If you are not sure, pick 1.0 or 1.1. If your span is too long for the size of lumber you need to use, move up to a stronger grade of lumber until you find one that works.

Reference Tables Codes specify the minimum spans for various horizontal framing members and the number and placement of nails for framing connections.

These tables are typical of most model codes, but check with your local building officials to make sure the figures are in compliance with the tables they use.

Typical Lumber Strengths

Balsam Fir	No. 1	E=1.2 F_b=1150–1440
	No. 2	E=1.1 F_b=950–1190
California Redwood	No. 1	E=1.1 F_b=1350–1690
	No. 2	E=1.0 F_b=1100–1370
Douglas Fir (Larch)	No. 1	E=1.8 F_b=1750–2190
	No. 2	E=1.7 F_b=1450–1810
Douglas Fir (South)	No. 2	E=1.3 F_b=1350–1690
Northern Pine	No. 1	E=1.4 F_b=1400–1750
	No. 2	E=1.3 F_b=1100–1380
Lodgepole Pine	No. 1	E=1.5 F_b=1300–1620
	No. 2	E=1.3 F_b=1050–1310
Southern Pine	No. 2	E=1.6 F_b=1400–1750

Note: The symbol E stands for Modulus of Elasticity, a measurement of tensile strength; F_b stands for Fiber Stress in Bending.

Floor Joists

Allowable spans for 40 pounds per square foot live load. The figures labeled OC mean inches on center.

Joist Size	Joist Spacing	Modulus of Elasticity, E, in 1,000,000 psi													
		0.8	0.9	1.0	1.1	1.2	1.3	1.4	1.5	1.6	1.7	1.8	1.9	2.0	2.2
2×6	16" OC	7'9"	8'0"	8'4"	8'4"	8'10"	9'1"	9'4"	9'6"	9'9"	9'11"	10'2"	10'4"	10'6"	10'10"
	24" OC	6'9"	7'0"	7'3"	7'6"	7'9"	7'11"	8'2"	8'4"	8'6"	8'8"	8'10"	9'0"	9'2"	9'6"
2×8	16" OC	10'2"	10'7"	11'0"	11'4"	11'8"	12'0"	12'3"	12'7"	12'10"	13'1"	13'4"	13'7"	13'10"	14'3"
	24" OC	8'11"	9'3"	9'7"	9'11"	10'2"	10'6"	10'9"	11'0"	11'3"	11'5"	11'8"	11'11"	12'1"	12'6"
2×10	16" OC	13'0"	13'6"	14'0"	14'6"	14'11"	15'3"	15'8"	16'0"	16'5"	16'9"	17'0"	17'4"	17'8"	18'3"
	24" OC	11'4"	11'10"	12'3"	12'8"	13'0"	13'4"	13'8"	14'0"	14'4"	14'7"	14'11"	15'2"	15'5"	15'11"
2×12	16" OC	15'10"	16'5"	17'0"	17'7"	18'1"	18'7"	19'1"	19'6"	19'11"	20'4"	20'9"	21'1"	21'6"	22'2"
	24" OC	13'10"	14'4"	14'11"	15'4"	15'10"	16'3"	16'8"	17'0"	17'5"	17'9"	18'1"	18'5"	18'9"	19'4"

Ceiling Joists

Allowable spans for 10 pounds per square foot live load (wallboard ceiling).

Joist Size	Joist Spacing	0.8	0.9	1.0	1.1	1.2	1.3	1.4	1.5	1.6	1.7	1.8	1.9	2.0	2.2
2×4	16" OC	8'11"	9'4"	9'8"	9'11"	10'3"	10'6"	10'9"	11'0"	11'3"	11'6"	11'9"	11'11"	12'2"	12'6"
	24" OC	7'10"	8'1"	8'5"	8'8"	8'11"	9'2"	9'5"	9'8"	9'10"	10'0"	10'3"	10'5"	10'7"	10'11"
2×6	16" OC	14'1"	14'7"	15'2"	15'7"	16'1"	16'6"	16'11"	17'4"	17'8"	18'1"	18'5"	18'9"	19'1"	19'8"
	24" OC	12'3"	12'9"	13'3"	13'8"	14'1"	14'5"	14'9"	15'2"	15'6"	15'9"	16'1"	16'4"	16'8"	17'2"
2×8	16" OC	18'6"	19'3"	19'11"	20'7"	21'2"	21'9"	22'4"	22'10"	23'4"	23'10"	24'3"	24'8"	25'2"	25'11"
	24" OC	16'2"	16'10"	17'5"	18'0"	18'6"	19'0"	19'6"	19'11"	20'5"	20'10"	21'2"	21'7"	21'11"	22'8"
2×10	16" OC	23'8"	24'7"	25'5"	26'3"	27'1"	27'9"	28'6"	29'2"	29'9"	30'5"	31'0"	31'6"	32'6"	33'1"
	24" OC	20'8"	21'6"	22'3"	22'11"	23'8"	24'3"	24'10"	25'5"	26'0"	26'6"	27'1"	27'6"	28'0"	28'11"

Low Sloped Rafters

Allowable spans for roofs 3 in 12 or less (or any high slope rafter); 30 pounds per square foot live load (supporting drywall ceiling and heavy roof covering).

Rafter Size	Rafter Spacing	Allowable Extreme Fiber Stress in Bending F_b psi														
		500	600	700	800	900	1000	1100	1200	1300	1400	1500	1600	1700	1800	1900
2×6	16" OC	6'6"	7'1"	7'8"	8'2"	8'8"	9'2"	9'7"	10'0"	10'5"	10'10"	11'3"	11'7"	11'11"	12'4"	12'8"
	24" OC	5'4"	5'10"	6'3"	6'8"	7'1"	7'6"	7'10"	8'2"	8'6"	8'10"	9'2"	9'6"	9'9"	10'0"	10'4"
2×8	16" OC	8'7"	9'4"	10'1"	10'10"	11'6"	12'1"	12'8"	13'3"	13'9"	14'4"	14'10"	15'3"	15'9"	16'3"	16'8"
	24" OC	7'0"	7'8"	8'3"	8'10"	9'4"	9'10"	10'4"	10'10"	11'3"	11'8"	12'1"	12'6"	12'10"	13'3"	13'7"
2×10	16" OC	10'11"	11'11"	12'11"	13'9"	14'8"	15'5"	16'2"	16'11"	17'7"	18'3"	18'11"	19'6"	20'1"	20'8"	21'3"
	24" OC	8'11"	9'9"	10'6"	11'3"	11'11"	12'7"	13'2"	13'9"	14'4"	14'11"	15'5"	15'11"	16'5"	16'11"	17'4"
2×12	16" OC	13'3"	14'6"	15'8"	16'9"	17'9"	18'9"	19'8"	20'6"	21'5"	22'2"	23'0"	23'9"	24'5"	25'2"	25'10"
	24" OC	10'10"	11'10"	12'10"	13'8"	14'6"	15'4"	16'1"	16'9"	17'5"	18'1"	18'9"	19'4"	20'0"	20'6"	21'1"

High Slope Rafters

Allowable spans only for slope greater than 3 in 12; 30 pounds per square foot live load (supporting drywall ceiling and heavy roof covering).

Rafter Size	Rafter Spacing	500	600	700	800	900	1000	1100	1200	1300	1400	1500	1600	1700	1800	1900
2×4	16" OC	4'1"	4'6"	4'11"	5'3"	5'6"	5'10"	6'1"	6'5"	6'8"	6'11"	7'2"	7'5"	7'7"	7'10"	8'0"
	24" OC	3'4"	3'8"	4'0"	4'3"	4'6"	4'9"	5'0"	5'3"	5'5"	5'8"	5'10"	6'0"	6'3"	6'5"	6'7"
2×6	16" OC	6'6"	7'1"	7'8"	8'2"	8'8"	9'2"	9'7"	10'0"	10'5"	10'10"	11'3"	11'7"	11'11"	12'4"	12'8"
	24" OC	5'4"	5'10"	6'3"	6'8"	7'1"	7'6"	7'10"	8'2"	8'6"	8'10"	9'2"	9'6"	9'9"	10'0"	10'4"
2×8	16" OC	8'7"	9'4"	10'1"	10'10"	11'6"	12'1"	12'8"	13'3"	13'9"	14'4"	14'10"	15'3"	15'9"	16'3"	16'8"
	24" OC	7'0"	7'8"	8'3"	8'10"	9'4"	9'10"	10'4"	10'10"	11'3"	11'8"	12'1"	12'6"	12'10"	13'3"	13'7"
2×10	16" OC	10'11"	11'11"	12'11"	13'9"	14'8"	15'5"	16'2"	16'11"	17'7"	18'3"	18'11"	19'6"	20'1"	20'8"	21'3"
	24" OC	8'11"	9'9"	10'6"	11'3"	11'11"	12'7"	13'2"	13'9"	14'4"	13'11"	15'5"	15'11"	16'5"	16'11"	17'4"

INDEX

U.S./Metric Measure Conversion Chart

	Symbol	When you know:	Multiply by:	To find:	Rounded Measures for Quick Reference		
Formulas for Exact Measures							
Mass (weight)	oz	ounces	28.35	grams	1 oz		= 30 g
	lb	pounds	0.45	kilograms	4 oz		= 115 g
	g	grams	0.035	ounces	8 oz		= 225 g
	kg	kilograms	2.2	pounds	16 oz	= 1 lb	= 450 g
					32 oz	= 2 lb	= 900 g
					36 oz	= 2¼ lb	= 1000 g (1 kg)
Volume	pt	pints	0.47	liters	1 c	= 8 oz	= 250 ml
	qt	quarts	0.95	liters	2 c (1 pt)	= 16 oz	= 500 ml
	gal	gallons	3.785	liters	4 c (1 qt)	= 32 oz	= 1 liter
	ml	milliliters	0.034	fluid ounces	4 qt (1 gal)	= 128 oz	= 3¾ liter
Length	in.	inches	2.54	centimeters	⅜ in.	= 1.0 cm	
	ft	feet	30.48	centimeters	1 in.	= 2.5 cm	
	yd	yards	0.9144	meters	2 in.	= 5.0 cm	
	mi	miles	1.609	kilometers	2½ in.	= 6.5 cm	
	km	kilometers	0.621	miles	12 in. (1 ft)	= 30 cm	
	m	meters	1.094	yards	1 yd	= 90 cm	
	cm	centimeters	0.39	inches	100 ft	= 30 m	
					1 mi	= 1.6 km	
Temperature	°F	Fahrenheit	⅝ (after subtracting 32)	Celsius	32° F	= 0° C	
	°C	Celsius	⁹⁄₅ (then add 32)	Fahrenheit	68° F	= 20° C	
					212° F	= 100° C	
Area	in.²	square inches	6.452	square centimeters	1 in.²	= 6.5 cm²	
	ft²	square feet	929.0	square centimeters	1 ft²	= 930 cm²	
	yd²	square yards	8361.0	square centimeters	1 yd²	= 8360 cm²	
	a.	acres	0.4047	hectares	1 a.	= 4050 m²	